Our House is on Fire

Our House is on Fire

BEATA *and* MALENA
ERNMAN
GRETA *and* SVANTE
THUNBERG

ALLEN LANE
an imprint of
PENGUIN BOOKS

ALLEN LANE

UK | USA | Canada | Ireland | Australia
India | New Zealand | South Africa

Allen Lane is part of the Penguin Random House group of companies
whose addresses can be found at global.penguinrandomhouse.com

First published in Sweden as *Scener ur Hjärtat* by Bokförlaget Polaris 2018
This translation first published 2020
001

Set in 12/14.75 pt Dante MT Std
Typeset by Jouve (UK), Milton Keynes
Printed and bound in Great Britain by Clays Ltd, Elcograf S.p.A.

A CIP catalogue record for this book is available from the British Library

Hardback ISBN: 978–0–241–44673–7
Trade Paperback ISBN: 978–0–241–44674–4

www.greenpenguin.co.uk

Our House is on Fire

This could have been my story. An autobiography of sorts, had I been so inclined.

But autobiographies don't really interest me.

There are other more important things.

This story was written by Svante and me together with our daughters, and it's about the crisis that struck our family.

It's about Greta and Beata.

But above all it's about the crisis that surrounds and affects us all. The one we humans have created through our way of life: beyond sustainability, divorced from nature, to which we all belong. Some call it over-consumption, others call it a climate crisis.

The vast majority seem to think that this crisis is happening somewhere far away from here, and that it won't affect us for a very long time yet.

But that's not true.

Because it's already here and it's happening around us all the time, in so many different ways. At the breakfast table, in school corridors, along streets, in houses and apartments. In the trees outside your window, in the wind that ruffles your hair.

Perhaps some of the things that Svante and I, along with the children, decided to share here, after considerable deliberation, should have been saved for later.

Once we had more distance.

Not for our sake, but for yours.

No doubt this would have been perceived as more acceptable. A bit more agreeable.

But we don't have that kind of time. To have a fighting chance, we have to put this crisis in the spotlight right now.

A few days before this book was first published in Sweden in August 2018, our daughter Greta Thunberg sat down outside the Swedish Parliament and began a school strike for the climate – a strike that is still going on today, on Mynttorget in the Old Town in Stockholm, and in many places around the world.

Since then a lot has changed. Both for her and for us as a family.

Some days it's almost like a fairy tale. A saga.

But that's a story for another book.

This story is about the road to Greta's school strike. The road to 20 August 2018.

Malena Ernman, November 2018

P.S. Before this book was published we announced that any money we might earn from it would go to Greenpeace, the World Wildlife Fund, and other non-profit organizations, through a foundation we've set up.

And that's how it is.

Because that's what Greta and Beata have decided.

I

Behind the Curtain

Elegy

For the day wears on.
The sun will die at seven.
Speak up, experts on darkness,
who will brighten us now?
Who turns on a Western backlight,
who dreams an Eastern dream?
Someone, anyone, bring a lantern!
Preferably you.

– Werner Aspenström

Scene 1. One Last Night at the Opera

It's places, everyone.

The orchestra tune their instruments one last time and the lights go down in the hall. I'm standing next to the conductor, Jean-Christophe Spinosi, we're just about to walk through the stage door and take our positions.

Everyone is happy tonight. It's the final performance, and tomorrow we all get to go home to our loved ones. Or on to the next job. Home to France, Italy and Spain. Home to Oslo and Copenhagen. On to Berlin, London and New York.

The last few performances have felt like being in a trance.

Anyone who has ever worked on stage knows what I mean. Sometimes there is a kind of flow, an energy that builds in the interaction between stage and audience and sets off a chain reaction that unfolds from performance to performance, from night to night. It's like magic. Theatre and opera magic.

And now we're at the final performance of Handel's *Xerxes* at the Artipelag arts centre in the Stockholm archipelago. It is 2 November 2014, and on this evening I will sing my last opera in Sweden. But no one is aware of that. Including me.

This evening I will sing my last opera ever.

The atmosphere is electric, and everyone backstage is walking on air, a few centimetres above Artipelag's brand new concrete floor.

They are filming as well. Eight cameras and a full-scale production team are recording the performance.

Through the stage door you can hear the sound of 900 silent people. The King and Queen are in attendance. Everyone is there.

I'm pacing back and forth. I'm trying to breathe, but I can't. My body seems to want to twist to the left and I'm sweating. My hands are falling asleep. The last seven weeks have been one long nightmare. Nowhere is there the slightest bit of calm. I feel sick, yet beyond nausea. Like a drawn-out panic attack.

As if I had slammed right into a glass wall and got stuck mid-air as I was falling to the ground. I'm waiting for the thud. Waiting for the pain. I'm waiting for blood, broken bones and the wail of ambulances.

But nothing happens. All I see is myself suspended in the air in front of that bloody glass wall, which just stands there without the slightest crack.

'I'm not feeling well,' I say.

'Sit down. Do you want some water?' We're speaking French, the conductor and I.

Suddenly my legs give out. Jean-Christophe catches me in his arms.

'It's fine,' he says. 'We'll delay the performance. They can wait. We'll blame it on me, I'm French. We're always late.'

Someone laughs.

I really have to hurry home after the performance. My younger daughter, Beata, is turning nine tomorrow and I have a thousand things to take care of. But now I am where I am. Unconscious, in the arms of the conductor.

Typical.

Someone caresses my forehead.

Cut to black.

Scene 2. The Ironworks

I grew up in a terrace house in the small, northern town of Sandviken, Sweden. Mum was a deacon and Dad worked as an accountant at the Sandvik ironworks. I have a sister, Vendela, three years my junior, and a brother, Karl-Johan. Mum named him after the Swedish baritone Carl-Johan 'Loa' Falkman because she thought Loa was so handsome.

This was the extent of the connection to opera and classical music I had at home.

We did like to sing though. We sang a lot. Folk music, ABBA, John Denver. All in all we were just another small-town Swedish family. The only thing that might have set us apart was my parents' involvement with vulnerable people.

In our little home on Ekostigen, on the outskirts of the Vallhov neighbourhood, humanitarianism reigned – if someone needed help it was our duty to try to offer that help to them. My mother carried on this family tradition from her father, Ebbe Arvidsson, who was a high-up official in the Church of Sweden and a pioneer in ecumenism and modern aid work. So in my younger years I often found myself living under the same roof as refugees and undocumented immigrants.

It could be a bit chaotic at times.

But it worked out fine.

Whenever we travelled somewhere it was always to visit my mother's best friend, who was a nun; we spent several summers at her convent in the north of England. This is probably why I swear so much on stage. The habit must stem from a sort of chronic childhood tendency towards rebellion, which I don't think I'll ever quite shake.

But apart from the fact that we spent our summers in the dormitories of English convents and that we had refugees living in the garage, we were just like everyone else.

As I said, we sang, and I loved to sing. I sang all the time.

I sang everything and anything – the harder the piece, the more fun it was.

Much later, when I chose to become an opera singer, it was probably for the simple fact that I love a challenge. After all, nothing was harder or more fun than singing opera.

Scene 3. Cultural Workers

Since I was six years old, I've been singing on stage for audiences. Church choirs, vocal groups, jazz bands, musicals, opera. My love for song is boundless – I would rather not stick to any one genre or be put in any fach. My work sprawls in every direction and across all boundaries. I will sing anything as long as the music is good.

In the entertainment industry it's often said that the easier you are to define as an artist the more cookbooks you'll get to write – and my cookbooks presumably shine brighter in their absence than most.

But the past fifteen years, in my view anyway, have a clear through-line, in that I've been trying to combine the height of artistry with a broad, popular repertoire. I've tried to simplify the difficult, take high culture down a notch, widen what is narrow. And vice versa.

I've gone my own way. Against the current and most often alone. Except when Svante was involved, obviously.

What was at first driven by instinct and intuition became a method over the years – akin to a responsibility, a conviction that if you have the ability to push the boundaries of your profession, you are also obliged to try.

Svante and I are among the very few who were given that opportunity.

And we tried.

We are 'cultural workers' – a term widely used in Sweden for anyone working within the arts. We are trained in opera, music and theatre, with half a career of freelance and institutional work behind us. We do what all cultural workers are ultimately programmed to do. We strive and toil to secure our future and reach our eternal goal: finding that new audience.

We come from quite different backgrounds, but have shared the same goal, right from the start.

Different, but similar.

When I was pregnant with our first child, Greta, and working in Germany, Svante was working at three different theatres in Sweden: Östgötateatern, Riksteatern and Orionteatern. Simultaneously. I had several years of binding contracts ahead of me at various opera houses all over Europe. With a thousand kilometres between us, we talked over the phone about how we could get our new reality to work.

'You're one of the best in the world at what you do,' Svante said. 'I've read that in at least ten different newspapers, and as for me, it's more like I'm a bass player in the Swedish theatre. I can very easily be replaced. Not to mention the fact you earn so damned much more than me.'

'Than *I do*.'

'You earn so damned much more than *I do*.'

I protested a little half-heartedly but the choice was made, and after Svante's last performance he flew down to join me in Berlin.

The next day Svante's phone rang. He took the call standing on a balcony overlooking Friedrichstrasse. It was late May and the summer heat was already oppressive. We hadn't even been together for six months.

'So bloody typical,' he said, laughing, after he'd hung up a few minutes later.

'Who was that?'

'It was Erik Haag and another guy. They were at the Orion and saw the performance last week.'

Svante had acted with Helena af Sandeberg in a play by Irvine Welsh, author of *Trainspotting*; everyone was doing drugs and swaddling corpses in shrink wrap.

'Fuck me!' was a line that Helena had screamed at Svante several evenings a week since the play opened.

Helena was considered to be the best-looking actress in all of Sweden and I was terribly jealous about the whole thing.

'They're doing a comedy programme for one of the big networks and think I seem like a funny guy, so they were wondering if I wanted to be on it. This is exactly the kind of call I've been waiting for . . .'

'What did you say? You have to do it!' I fixed my eyes on him.

'I told them I'm with my very pregnant girlfriend and that she is working abroad.' He fixed his eyes on me.

'You turned it down?'

'Yes. It's the only way. We're doing this together, otherwise it'll never work.'

And so it was.

A few weeks later we were sitting at the premiere party for the *Don Giovanni* production I was singing in at the Berlin Staatsoper and Svante explained his current professional status to my colleagues, Maestro Barenboim and Cecilia Bartoli.

'So now I'm a housewife.'

Then we kept on doing that for twelve years. It was arduous and hard but also great fun. We spent two months in each city and then we moved on to the next one. Berlin, Paris, Vienna, Amsterdam, Barcelona. Round and round.

We spent the summers in Glyndebourne, Salzburg or Aix en Provence. As you do when you're good at singing opera and other classical music.

I rehearsed twenty to thirty hours a week and the rest of the time we spent together. Free. No relatives, except Grandmother Mona, who sometimes came to stay. No friends. No dinners. No parties. Just us.

Beata was born three years after Greta and we bought a Volvo V70 so we'd have room for doll's houses, teddy bears and tricycles. We carried on. Round and round and round. Those were fantastic years. In wintertime we'd sit in bright, beautiful fin-de-siècle apartments, playing with the girls on the floor, and when spring came we'd stroll out through green, leafy parks together.

Our everyday life was like no one else's. Our everyday life was absolutely marvellous.

Scene 4.
Unique Opportunities

'Being involved in Melodifestivalen [the annual song competition in Sweden ahead of the Eurovision Song Contest] is a little like having a baby. You can tell others about it, you can describe it in detail. But only those who've been there can know how it feels.'

Anders Hansson is a music producer and we're about to start working on my next album. At the moment we're dragging our suitcases across Stortorget in Malmö on our way to the Stockholm train and Anders is laughing as he explains the situation to Svante and me.

It's the morning after my schlager debut. A big picture of me, Petra Mede and Sarah Dawn Finer adorns the front page of *Aftonbladet*. The caption reads: 'Malmö Arena at 9.23 p.m.' In the picture I am radiating total and utter shock.

Melodifestivalen is the biggest annual TV show in all the Nordic countries. It lasts for six weeks and can only be described as a happy mix between the *X Factor* and the American Super Bowl. Almost half of the population of Sweden watch the final show, and the media surrounding it is second to none. If you enter, you must be prepared for a bumpy ride . . . especially if you do it as an opera singer, like me. But I decided, if you enter Melodifestivalen, you enter to win, and to win well. In other words, the odds should be on you coming in last. You should be alongside all the major performers in the finale – and then you should win by the smallest conceivable margin, preferably decided by the people's vote. So that's exactly what I did. It was the biggest upset in the history of the whole competition.

After that, you'd think it would just be a matter of using our new platform and getting to work. That 'new audience' had indeed been found and the groundwork could not have been laid better.

Melodifestivalen gave us a unique chance – an opportunity that would probably never come again. The audiences poured in. The Minister of Culture talked about a 'Malena effect'.

'Opera's being led away from the salons and back to the people,' was the headline in *Expressen*. And *Dagens Nyheter*'s culture editor wrote: 'It's too good to be true. But it's true.'

For a brief moment I almost believed it was possible: opera could once again be made broad public entertainment.

But when autumn arrived everything was back to normal. Swedish opera institutions weren't calling wanting to make the most of the opportunity. The audience was there for it, but no one seemed interested in reaching them.

So we decided to do everything ourselves.

Title roles in opera houses abroad and a solo artist at home, self-produced concerts, tours and performances.

All in our pursuit of the new, wider audience.

One evening two weeks before the final *Xerxes* performance, Svante and I sat slumped on our bathroom floor in Stockholm. It was late, the children were asleep. Everything was starting to fall apart around us. Our apartment walls were behaving differently. Huge cracks had started running across the floor and ceiling and it felt like the whole block would at any second give way and slide down into Lake Klara.

Greta was eleven, had just started fifth grade, and was not doing well. She cried at night when she should have been sleeping. She cried on her way to school. She cried in her classes and during her breaks, and the teachers called home almost every day. Svante had to run off and bring her home. Home to Moses, because only Moses offered any functioning help.

She sat with our Golden Retriever for hours, petting him and stroking his fur. We tried our best, but nothing helped. She was slowly disappearing into some kind of darkness and little by little, bit by bit, she seemed to stop functioning. She stopped playing the piano. She stopped laughing. She stopped talking.

And.

She stopped eating.

We sat there on the hard mosaic floor, knowing exactly what we would have to do. We would have to do everything. We would have to change everything. We would have to find the way back to Greta, no matter the cost.

But that would not be enough. The situation called for more than words and feelings. A closing of accounts. A clean break.

'How are you feeling?' Svante asked. 'Do you want to keep going?'

'No.'

'Okay. Fuck this. No more,' he said. 'You can't make opera popular when the opera institutions don't want opera to be popular. And it doesn't matter if someone else finds that *new* audience when no one seems to want it.'

'I agree. I'm done.' And I was.

'If it's not enough getting twenty thousand people to drive through the woods to an art gallery on an island, three kilometres from the nearest bus stop, all with no sponsors and not a single penny in subsidy . . . if even *that* is not enough, not a goddamn thing is going to be enough.'

Svante has a temper, which is not always to his advantage. But there wasn't much to object to in his conclusion.

'We've taken it as far as we can,' I said. 'I honestly don't think I'd survive if we continued.'

'So we'll cancel everything. Every last contract,' Svante went on. 'Madrid, Zurich, Vienna, Brussels. Everything. We'll find something to blame it on. Then we'll change tack. Concerts, musicals, theatre, TV. Sing opera. Sing the music, but no more opera performances.'

'I'll do the final show in two weeks. Then no more.'

I'd made my decision.

'Should we announce it? Or would that be stupid?'

'Yes,' I said. 'That would probably be a stupid thing to do.'

So we didn't say anything.

Scene 5.
Xerxes: King of Persia

It turns out that I was unconscious for almost ten minutes. The audience was informed that unfortunately the performance would be delayed by a few minutes.

Behind the curtain, there was a buzz of discussion about how the situation should be handled, but that was none of my concern. I knew exactly what I was going to do.

It was time to end this once and for all.

I took a sip of water and nodded at the conductor.

'Can you stand up?'

'No.' I stood up.

'Can you walk?'

'No.' I walked towards the stage door. Worried looks flitted all around me.

'But can you sing?'

'No,' I said, nodding at the stage manager, as I strode out onto the stage.

Those who were there say the applause that night was something special. People stood up and cheered in a way they don't usually do.

Everyone backstage was carried by the wings of intoxication. Like in a movie. The King and Queen gave an ovation, and it was as if everyone was speaking through laughter.

As if it were all in slow motion. At half the speed.

Pernilla, my agent, helped me off with my costume and wig. 'Don't tell Svante what happened. He'll only worry for no reason.'

She nodded.

From above came voices from the lobby: Swedish, French, German, Spanish.

They all sounded so happy. And as I was being carried out to the taxi, I saw them raise their champagne glasses in a toast. Three cheers and hip-hip-hurrah.

I lay down on the back seat and cried the whole way into the city.

Not because I was sad. Not because I was relieved. Not because everything was what it was.

I was crying because I had no memory of the performance.

It was as though I hadn't been there at all.

Scene 6. Gnocchi

Breakfast: ⅓ banana. Time: 53 minutes

On a white sheet of paper fixed to the wall we note down everything Greta eats and how long it takes for her to eat it. The amounts are small. And it takes a long time. But the emergency unit at the Stockholm Centre for Eating Disorders says that this method has a good long-term success rate. You write down what you eat meal by meal, then you list everything you can eat, things you wish you could eat and things you want to be able to eat further down the line.

It's a short list.

Rice, avocados and gnocchi.

It's Tuesday 11 November 2014 and we find ourselves somewhere between the abyss and Kungsholms Strand in Stockholm. School starts in five minutes. But there isn't going to be any school today. There isn't going to be any school at all this week.

Yesterday Svante and I got another email from the school expressing their 'concern' about Greta's lack of attendance, despite the fact that they were in possession of several letters from both doctors and psychologists explaining her situation.

Again, I inform the school office of our situation and they reply with an email saying that they hope Greta will come to school as usual on Monday so 'this problem' can be dealt with.

But Greta won't be in school on Monday. Because Greta stopped eating two months ago and unless a sudden dramatic change occurs she's going to be admitted to Sachsska Children's Hospital next week.

We have lunch on the sofa in front of *Once Upon a Time* on DVD. There are several seasons and each season lasts approximately half a geologic age. Fitting. We need oceans of time to get us through our meals.

Svante is boiling gnocchi. It is extremely important that the consistency is perfect, otherwise it won't be eaten. All of this over gnocchi, small pasta dumplings made of potato and shaped like little rugby balls.

We set a specific number of gnocchi on her plate. It's a delicate balancing act; if we offer too many our daughter won't eat anything and if we offer too few she won't get enough. Whatever she ingests is obviously too little, but every little bite counts and we can't afford to waste a single one.

Then Greta sits there sorting the gnocchi. She turns each one over, presses on them and then does it again. And again. After twenty minutes she starts eating. She licks and sucks and chews: tiny, tiny bites. It takes forever. An episode ends. Thirty-nine minutes. We start the next one and note the time in between bites, the number of bites per episode, but we keep quiet.

'I'm full,' she says suddenly. 'I can't eat any more.'

Svante and I avoid looking at each other. We have to hold back our frustration, because we've started to realize that this is the only thing that works. We've explored all other tactics. Every other conceivable way.

We've ordered her sternly. We've screamed, laughed, threatened, begged, pleaded, cried and offered every imaginable bribe. But this seems to be what works the best.

Svante goes up to the sheet of paper on the wall and writes: *Lunch: 5 gnocchi. Time: 2 hours and 10 minutes.*

Scene 7. On the Art of Baking Cinnamon Buns

It's the third weekend in September 2014, and later in the afternoon I'm going to Artipelag. But now it's all about cinnamon buns.

We're going to bake buns, all four of us, the whole family, and we're determined to make this work. It has to.

If we can bake our buns as usual, in peace and quiet, Greta will be able to eat them as usual, and then everything will be resolved, fixed. It's going to be easy as pie. Baking buns is after all our favourite activity.

So we bake, dancing around in the kitchen so as to create the most positive, happiest bun-baking party in human history.

But once the buns are out of the oven the party stops in its tracks. Greta picks up a bun and sniffs it. She sits there holding it, tries to open her mouth, but . . . can't. We see that this isn't going to work.

'Please eat,' Svante and I say in chorus.

Calmly, at first.

And then more firmly.

Then with every ounce of pent-up frustration and powerlessness.

Until finally we scream, letting out all our fear and hopelessness. 'Eat! You have to eat, don't you understand? You have to eat now, otherwise you'll die!'

Then Greta has her first panic attack. She makes a sound we've never heard before, ever. She lets out an abysmal howl that lasts for over forty minutes. We haven't heard her scream since she was an infant.

I cradle her in my arms, and Moses lies alongside her, his moist nose pressed to her head.

The buns are in a heap on the kitchen floor.

After an hour she is calm, and we assure her that we aren't going to eat any more buns, not to worry.

'Everything will work out, everything will be fine.'

Then it's time for me to go to the performance. It's a matinée. The family accompanies me to Artipelag and in the car Greta asks, 'Am I going to get well again?'

'Of course you are,' I reply.

'When am I going to get well?'

'I don't know. Soon.'

The car stops outside the spectacular building.

I go backstage and start warming up my voice.

Scene 8. At the Children's Hospital

No matter how bad I've felt in my life, I've always felt good on stage. The stage is my sanctuary. But now some kind of boundary has been crossed and each performance of *Xerxes* is utter darkness. I don't want to stand there. I don't want to be there. I want to be at home with my children. I want to be anywhere but at the bloody, fucking Artipelag.

And most of all I want to be able to answer Greta's question: 'When am I going to get well?'

But I don't have an answer to that. No one has an answer because first of all we have to figure out which illness she has.

It all starts at the health clinic a month into the autumn term. A couple of weeks have passed since we started noticing that something is not right, and a few days after Greta's blood tests we get a call from a young doctor.

The test results don't look so good, she says, and recommends that we go to Astrid Lindgren Children's Hospital to have additional, more thorough tests done.

'Should we schedule an appointment?' Svante asks.

'No,' the doctor replies. 'I think you should go there right now.' Fifteen minutes later we've picked Greta up from school and are on our way to the Emergency Room. There we continue taking more tests and samples, and then we have to sit down and wait.

So we wait. As the pressure and worry grow. We call Svante's mother, and ask her to pick Beata up from school.

Several hours later, a new doctor comes over and talks to

us. Certain results indicate that something is seriously wrong. They're just not sure exactly what yet. Svante collapses in a heap on the floor and for a few hours we find ourselves in freefall.

The gates to hell crack open and we pace back and forth in the tiny examination room where so many have paced before us and so many others will pace later.

We had bought a plastic-wrapped baguette with a curry filling; it's on the stainless-steel stool by the door. I'm sitting on the floor with Greta on my lap, trying to amuse her.

In the years that have passed since, we have often thought back to those hours. Although never in detail. Svante recalls his legs giving way in the corridor and I remember the heavy, boundless darkness that enveloped us and all of the other families waiting there, each in our own small consultation room. I only remember the little that I have chosen to remember. The rest I can't bear to think about.

Grazing that memory, even for a split second, is usually enough to put life in perspective.

Yet another doctor comes in. She moves the plastic-wrapped baguette off the stool and sits down, goes through the results and calms us. They've checked and everything looks fine. There are no signs that anything is wrong and we can breathe out, thank the gods and go home.

Taking the stage that night was not much fun, but it was, to say the least, a luxury compared with being one of the families that didn't get to leave the hospital that late afternoon; the families still sitting there in the consultation rooms facing the gates of hell.

A few days later we get another call from Astrid Lindgren Children's Hospital. The doctor recommends that we contact BUP, the psychiatric service for children and adolescents. The tests haven't shown anything that can't be explained by the fact that Greta has started having significant problems with her appetite.

'It's not unusual among girls in early puberty,' she says. 'Very often the cause is psychological rather than medical.'

Scene 9. Starvation

Sometimes the body's wisdom surpasses our own. Sometimes we use the body to express what we can't express in any other way. And sometimes, when we don't have the strength or the words to describe how we're feeling, we use the body as an interpreter.

Not eating can mean many things.

The question is what.

The question is why.

The thought that Svante or I would have managed to eat that plastic-wrapped curry baguette in the waiting room at Astrid Lindgren Children's Hospital is inconceivable, and the insight that Greta is feeling the way we did then, all the time, is somehow there, constantly grinding away at us.

Svante and I continue to look for answers. I spend the evenings reading everything I can find on the internet about anorexia and autism and eating disorders. We're sure it's not anorexia. But, we keep hearing that anorexia is a very cunning disorder and will do anything to evade discovery.

So we keep that door wide open.

Our life turns to chaos, and all forms of logic feel very far away. I read about hypersensitivity, gluten allergy, urinary tract infections, PANDAS and neuropsychiatric diagnoses.

During the day I sit with my phone and make calls from the moment I wake up until its too late to call anyone, stopping only to go out to Artipelag for the performances. All while Svante spends every waking minute trying to create a normal everyday life for the children.

I call the children's psychiatry service (BUP), the healthcare information service, doctors, psychologists and every conceivable superficial acquaintance who may be able to offer the least bit of knowledge or guidance. It is an endless chain of conversations, and 'I know someone who knows someone who knows someone . . .'

The adrenalin keeps me going for as long as it takes.

Despite the sleepless nights, and having lost all appetite, forgetting to eat.

My friend Kerstin knows Lina, who is a psychiatrist, and Lina spends hours talking to me. She listens, she offers advice, and she manages to get us an appointment at our local Kungsholmen BUP clinic.

At Greta's school there's a psychologist who is experienced with autism. She talks with both of us on the phone and says that a careful investigation must of course still be conducted, but in her eyes – and off the record – Greta shows clear signs of being on the autism spectrum.

'High-functioning Asperger's,' the school psychologist says.

We do our best to take in what she's telling us, which does sound extremely convincing, but coming to terms with the fact that our daughter could be autistic is a tremendous challenge. Not a single person in our circle of acquaintances reacts with anything other than a big 'Huh?!' when we test the autism theory out on them.

Greta does not have a single characteristic trait of autism or Asperger's to us. So either the school psychologist is crazy or else we have run into a gigantic gap in public awareness.

Then follows an endless stream of meetings, from BUP to the Stockholm Centre for Eating Disorders, where we repeat our story and explore our options. We talk away while Greta sits silently. She has stopped speaking with anyone except me, Svante and Beata. We take turns relaying the details.

Sometimes there are up to six people present at the meetings, and although everyone really wants to offer all the help they can, it's as if there's no help to be had.

Not yet, at least.

We're fumbling in the dark.

After two months of not eating Greta has lost almost 10 kg, which is a lot when you are rather small to begin with. Her body temperature is low and her pulse and blood pressure clearly indicate signs of starvation.

She no longer has the energy to take the stairs and her scores on the depression tests she takes are sky high. We explain to our daughter that we have to start preparing ourselves for a stay at the hospital, and we describe how it's possible to get nutrition and food without eating, with tubes and drips.

Scene 10.
We Call Him the Hans Rosling of Eating Disorders

In mid-November there's a big meeting at BUP. Three people from the Centre for Eating Disorders are also present.

Greta sits silently. As usual. I'm crying. As usual.

'If there are no developments after the weekend then we'll have to admit you to the hospital,' the doctor says.

On the stairs down to the lobby Greta turns round.

'I want to start eating again.'

'We'll have a banana when we get home,' Svante says.

'No. I want to start eating again like normal.'

All three of us burst into tears and we go home and Greta eats a whole green apple. But nothing more will go down. As it turns out, it's a little harder than you think to just start eating again.

But even though this makes Greta sad she doesn't panic. She has made up her mind and we keep trying, and at last we find a little path that we can follow through the thicket.

We take a few careful, trial steps and it works. Our legs hold up.

We inch forward.

We have rice, avocado, calcium tablets, bananas and time.

We take our time.

Unlimited time.

Svante stays home, never leaving the children's side. We listen to audiobooks, complete jigsaws, do homework, read about nature and natural science and note every meal on the paper on the wall.

Beata disappears into her room as soon as she comes home

from school. We hardly see her. She notices our worry and avoids us.

Together with Greta we plough through *A Faraway Island*, *Around the World in 80 Days* and *A Man Called Ove*.

The whole emigrant series by Vilhelm Moberg. Strindberg, Selma Lagerlöf, Mark Twain, Emily Brontë and the Stockholm series by Per Anders Fogelström.

One banana, 25 minutes. One avocado with 25g of rice, 30 minutes.

Outside the window the last leaves are falling from the trees. And we start on the long, long road back.

After another two months the weight loss has not only stopped – it has reversed and now points slowly, cautiously upwards. To the list we have added salmon and hash browns.

At the Centre for Eating Disorders we have an amazing doctor who makes notes about weight and pulse rates and explains all about nutrients and the body's building blocks during long educational lectures in his consultation room. And we start with sertraline, an antidepressant.

Greta is smart. She has a photographic memory and can rattle off all the world's capitals. She knows the capitals of all the territories too. If I ask 'Kerguelen Islands?' she answers 'Port-aux-Français.'

'Sri Lanka?'

'Sri Jayawardenapura Kotte.'

If I say 'Backwards?' then she answers just as fast. But backwards, of course. Svante likes to say that she's a better version of him, who thirty-five years ago devoted his childhood to collecting airline timetables and learning them by heart. She can rattle off the periodic table from memory in under a minute but it bothers her that she doesn't know how to pronounce some of the elements.

Greta's teacher instructs her in her spare time. Two hours a week, at breaks and between periods, in the library. In secret. It's enough to make Greta pass all her subjects in fifth grade.

Without that teacher nothing would have worked.

Nothing.

'I've seen too many highly sensitive, high-functioning girls fall apart. I've seen enough,' she says. 'I've reached my limit.'

When people fall apart it's hard to put them back together, and despite the fact that there is lots of will and knowledge to be found, the tools are often blunt and can be hopelessly ineffective.

There is help to be had within the framework of the system. For some. For those who fit into one of the few available templates or patterns. Greta is not one of them.

We have fought almost round the clock for several months before it sinks in that we have to do everything ourselves, and we are far from alone in that insight.

We are stuck in a Catch-22 between three different institutions and devote all our waking hours going to meetings about what can be done maybe, later on.

In any functioning social structure there should of course be an entity with sufficient resources to educate and inform society about mental illness and the various diagnoses. An authority that devoted itself to increasing awareness among teachers, parents and children about things we ought to know. Such an authority would probably be the most beneficial investment ever made in the history of modern society.

But no such entity exists.

What does exist is paediatric and adolescent psychiatry, where everyone is burned out from struggling with a constantly growing workload and where much of the time is spent putting out fires. What does exist is a school system where all pupils must function in exactly the same way and where over-worked teachers on a conveyor belt end up hitting the wall.

So you have to do everything yourself.

You have to educate yourself, you have to battle and fight.

And you need to be lucky as hell.

Scene 11.
'Children are Mean'

'Do they always look at you that way?'

'Don't know. Think so.'

Svante and Greta have been at the end-of-term ceremony where they tried to make themselves invisible at the back of classrooms, corridors and stairwells.

When students openly point and laugh at you – even though you're walking alongside your parent – then things have gone too far. Way too far.

Being bullied is terrible. But being bullied without understanding that you're being bullied – that's worse.

Back home in the kitchen, Svante explains to me what they've just experienced while Greta eats her rice and avocado.

I get so angry at what I hear that I could tear down half the street we live on with my bare hands, but our daughter – surprisingly – has a very different reaction. She's happy. Not relieved or calm, but happy. Exuberantly happy.

She then spends her entire Christmas break telling us about incidents and stories that are just terrible. It's like a movie montage featuring every imaginable clichéd bullying scenario. Every single one on the list is in there.

Stories about being pushed over in the playground, wrestled to the ground, or lured into strange places, the freeze out, the systematic shunning and the safe space in the girls' toilets where she sometimes manages to hide and cry before the break monitors force her out into the playground again.

For a full year, the stories keep coming.

Svante and I inform the school, but the school isn't

sympathetic. They don't believe her. Their understanding of the situation is different. It's Greta's own fault, the school thinks; several children have said repeatedly that Greta has behaved strangely and spoken too softly and never says hello. The latter they write in an email.

They write worse things than that, which is lucky for us, because when we report the school to the Swedish Schools Inspectorate we're on a firm footing and there's no doubt that the inspectorate will rule in our favour.

Greta's teacher continues to teach her in secret. The school administration repeatedly orders her to stop, finally threatening her with losing her job if she so much as talks to Greta or us. And so it goes. Week in and week out. Greta sneaks in and out of the school library and Svante waits outside in the car.

I explain that she'll have friends again, later. But her response is always the same.

'I don't want to have any friend. Friends are children and all children are mean.'

Greta sits down next to Moses.

'I can be your friend,' says Beata.

'It'll work out,' Svante says, writing on the paper on the wall: *1.5 avocado, 2 pieces of salmon plus rice, calcium tablet. Time: 37 minutes.*

Scene 12.
The Revenge of the Invisible Girl

Greta's pulse rate is getting stronger, the Centre for Eating Disorder's data shows, and finally the weight curve turns upwards strongly enough for a neuropsychiatric investigation to begin.

Our daughter has Asperger's, high-functioning autism and OCD, obsessive-compulsive disorder.

'We could formally diagnose her with selective mutism, too, but that often goes away on its own with time.'

We aren't surprised. Basically this was the conclusion we drew several months ago.

The school psychologist is present when we get the diagnosis at BUP, and we thank her for being frank with us from the very start.

On the way out, Beata calls to tell us she's having dinner with a friend, and I feel a sting of guilt. This is the first time in a long while that she won't be eating dinner alone, by herself in the upstairs bedroom. Soon we'll take care of you too, darling, I promise her in my mind, but first Greta has to get well.

Summer is coming, and we walk the whole way home. We almost don't even need to ration the burning of calories any more.

Scene 13.
'You are the strange ones. I'm the one who's normal'

– Thåström

What happened to our daughter can't be explained simply by a medical acronym, a diagnosis or dismissed as 'otherness'. In the end she simply couldn't reconcile the contradictions of modern life. Things simply just didn't add up.

We, who live in an age of historic abundance, who have access to shared resources far beyond all imagination, can't afford to help vulnerable people in flight from war and terror – people like you and me, but who have lost everything.

One day in school, Greta's class watches a film about how much rubbish there is in the oceans. An island of plastic, larger than Mexico, is floating around in the South Pacific. Greta cries throughout the film. Her classmates are also clearly moved. Before the lesson is over the teacher announces that on Monday there will be a substitute teaching the class, because she's going to a wedding over the weekend, in Connecticut, right outside of New York.

'Wow, lucky you,' the pupils say.

Out in the corridor the trash island off the coast of Chile is already forgotten. New iPhones are taken out of fur-trimmed down jackets, and everyone who has been to New York talks about how great it is, with all those shops, and Barcelona has amazing shopping too, and in Thailand everything is so cheap, and someone is going with her mother to Vietnam over the

Easter break, and Greta can't reconcile any of this with any of what they have just seen together. It doesn't add up and there is an unbearable feeling of loneliness and hopelessness that just won't go away. Had she not been different she would have been able to handle that feeling, just like the rest of us. She could have left it in that classroom. But she couldn't. She never could.

There are hamburgers for lunch, but she can't eat.

It's hot and crowded in the school cafeteria. The noise is almost ear-splitting and suddenly that greasy chunk of meat on the plate is no longer a piece of food. It's a ground-up muscle from a living being with feelings, awareness and a soul. The trash island has imprinted itself on her retinas.

She starts crying and wants to go home, but going home isn't an option because here in the school cafeteria you have to eat dead animals and talk about fashion, celebrities, make-up and mobile phones.

You're supposed to take a plate full of food, say that it's gross and poke at it just enough before tossing it all in the bin – without signalling either autism or anorexia or anything else that might indicate that there's something wrong with you. That you're different to the norm.

Greta has a diagnosis, but it doesn't rule out the fact that she's right and the rest of us have got it all wrong.

Because however much she tried she could not work out that equation that all the rest of us had already solved, the equation that was the ticket to a functioning everyday life.

She saw what the rest of us did not want to see.

She belonged to the tiny minority who could see our CO_2 emissions with their naked eye. Not literally of course, but still. She saw the invisible, colourless, scentless, soundless abyss that our generation has chosen to ignore; the greenhouse gases streaming out of our chimneys, hovering upwards with the winds and transforming the atmosphere into a gigantic, invisible garbage dump.

She was the child, we were the emperor.

And we were all naked.

Scene 14.
Something That is Just a Little Off

No parent would think twice about jumping in front of a running train to save their child. It's an instinct that no one denies.

But when that 'train' is actually coming, it's very rarely an actual speeding locomotive.

Nor is it usually as clear as the split second it takes to throw out your arms to catch someone falling.

It's just something that's a tiny bit askew. A little off. And it almost never resembles those rescue scenes we see in the movies.

The contours of a much bigger picture emerged in parallel with the bullying, diagnoses, depression and isolation. For us, that parallel picture developed so slowly that it almost went completely unnoticed. The fact that something in our everyday life was seriously wrong.

It actually wasn't even that hard to see. Just very uncomfortable.

And once we fixed our gaze on it, it was as if we couldn't stop looking. Because the insight that comes with this bigger picture suddenly fills your entire field of vision, it changes everything, and every fibre of your body tells you to look away but you can't, because it's your child, and there's nothing you wouldn't do for your child.

It took us four full years to grasp that image; the image of a skewed whole that would go on to change our lives completely.

Scene 15. Virtue-signalling Junkies

I was thirty-eight years old before I became a celebrity. Before I won the Melodifestivalen singing competition, I was well-known.

But being a *celebrity* is something different; a phenomenon, quite impossible to explain to someone who hasn't experienced it themselves.

'But what happens if she wins?' my then agent asked as we, with bowed heads and heavy sighs, reviewed the calendar for 2009 in the middle of January.

'She's an opera singer,' Svante said with a laugh. 'Obviously she's not going to win.'

The day after Melodifestivalen, Svante, I and four journalists from *Aftonbladet* and *Expressen* flew to Frankfurt, where I was rehearsing Rossini's *La Cenerentola*, which was to premiere five days later. It was all a bit messy.

My agent had to go around begging my employers for a little time off because not only had I unexpectedly made it to the finals but now I'd actually won the Melodifestivalen and I had to be in Moscow to represent Sweden in Eurovision in the middle of a season where I was otherwise supposed to be singing title roles in Frankfurt, Vienna and Stockholm.

'But will you manage?' my agent asked.

'I can manage anything,' I answered.

Svante and I never usually go to premieres, and we've never gone to any celebrity parties – or any other parties for that matter.

Being socially shy makes a person incredibly efficient; as soon as my concerts or performances are over I go straight home.

If I'm working in Stockholm I usually get away before the audience and remove my make-up on my bicycle ride home. If I absolutely must go to my own premiere parties I sneak off early.

Our children and work. That's all we can manage, Svante and I. Everything else has to be put aside. That's how we work, how we write.

We try to give voice to something more important than ourselves, and for us environmental and climate issues have become the ultimate example, and consequence, of the prevailing twisted world order.

Before Greta and Beata brought that to our attention we focused on other things. Human rights. Equal rights. Refugees. We weren't that concerned when it came to the environment. We thought it was being looked after. We were wrong. We thought we would solve everything with technological development. We were wrong. We were challenged by our children and in the end we ran out of arguments and now we're rapidly running out of time.

We find ourselves in the midst of an acute sustainability crisis, where global warming is one aspect; but if landslides in West Africa are one consequence of this crisis, drought in the Middle East another, and rising water levels for the island nations in the Pacific Ocean a third, then the crisis is expressing itself in our part of the world in the form of stress disorders, isolation and growing waiting lists within paediatric and adolescent psychiatry.

The planet is talking to us through our bar charts and diagrams. We see the animated graphics chewing up the ice in the Arctic north. Planet Earth is running a fever, but that fever – like any other fever – is only a symptom. In this case, the global heating is a symptom of a greater sustainability crisis, in which our lifestyles and our values ultimately threaten our future survival.

The sustainability crisis is what it all boils down to. It includes everything we do – from air pollution to economic structures, and it leads us to the core of humankind's state of health.

Scene 16.
The Antwerp Zoo

In the winter of 2010 we rent a rather seedy apartment on Rue du Fossé aux Loups in Brussels. Beata has just turned four, and we're going to spend one of my days off at the Antwerp zoo. A bottle of lice remover has exploded in the big suitcase on the flight to Brussels and now everything we own smells like lice shampoo. All the Pippi and Madicken DVDs are ruined and the whole stairwell stinks of Paranix.

We get up early, and it's not even nine o'clock by the time we're ready to go to catch the train at Brussels Midi station. There's only one thing left to do: Beata has to put on a pair of clean socks. She is extremely sensitive about many things and clothing is no exception.

'No! It feels dry,' she'll scream, and wriggle around on the floor in the hall because a sweater or a pair of trousers doesn't sit quite right. Sometimes we have to carry her out to the elevator and put her clothes on in the stroller, but there are days when not even that is possible, days when everything gets derailed and none of our usual everyday tricks work.

And this is of course completely unsustainable.

So today we're going to make a stand and set an example. We're going to fight fire with fire. She has to put on a fresh pair of socks before we leave. But Beata refuses.

After two hours we propose meeting halfway: taking off the old, dirty socks she has been wearing for almost a month.

She refuses.

So today we draw a line, we put our foot down. It is not our first time, but today Svante and I have all day and we do not intend to fold.

At two o'clock we leave the apartment and take the train up to Antwerp.

Beata still isn't wearing any socks in her shoes. She dangles her feet contentedly from the seat on the train.

Our duel is over and Beata is the sole victor.

'You make naughty little Lotta on Troublemaker Street seem like Mahatma Gandhi,' Svante says, laughing.

Beata smiles her most mischievous smile, and as always in that moment she is totally irresistible. You melt.

She is quite content.

We are going to the zoo.

Scene 17. Meltdown

It's called a meltdown. An outburst caused by feelings getting pent up, until they can no longer be processed within the realm of what we call reasonable behaviour.

One of Beata's first meltdowns was on Christmas Eve, a month or so before the outing to Antwerp Zoo.

She couldn't handle the anticipation and impressions, and exploded in a fit where she simply lost control and ended up in something that could only be described as emotional chaos.

Nothing could contain her, and it ended with us wrestling on the floor until I held her, calm, in my embrace.

'Don't you see what you're doing?' I sobbed in despair.

'Yeah.'

'But why are you doing it?'

Beata was crying too.

'I don't know.'

There were plenty of clues that something really wasn't right, but that didn't matter. To us, the logical thing to do was to scream and wave our arms around, demanding that a four-year-old could explain her own bad behaviour. Like two idiots.

'I think Beata has ADHD,' I said later to Svante. 'This isn't just ordinary disobedience.'

I don't know how I arrived at that conclusion right then, and even though today I know that our suspicions wouldn't have led to us getting any help for years, I still wish we'd followed that train of thought much further than we did.

Only deep in retrospect do we understand the extent of our

resistance to admitting that the situation was out of the ordinary. Instead, we blamed ourselves and adapted. As you do.

At pre-school Beata is a little angel, as she is everywhere outside the home. Clever, kind, shy and quite wonderfully charming. She is brilliant at playing the social game, and the slightest hint that we will tell her teachers about how she behaves at home causes her to break down.

Of course, we have no idea yet that these are all clear, early signs of ADHD in girls. Because how could we? There hasn't exactly been a public-awareness campaign.

We only know what we know. And we only do as we've been taught. To set limits and bring up your children to be functioning members of society. Then we continue.

So we continue, to tell her off. We continue to teach manners. Setting clear boundaries. And then we throw ourselves in the car, googling hotels on the phone as we drive north. Towards the Åre ski resort. Because the family doesn't function when it's just the four of us, Svante reasons, all we have to do is surround ourselves with other people at hotels and restaurants and then everything will be fine. Or better, at least, and 'It will work out later, you'll see'.

And Svante's rational thinking works! Apart from some sweat, stress and tears on the children's ski slope, we're doing fine again. We're functioning.

We learn how to ski. We drink hot chocolate and eat sausages with French fries for lunch.

In the afternoons we swim in the pool and then we dine out. It's lovely.

We have postponed the problem and swept everything under the rug in favour of a functioning school holiday and a little peace and quiet. We prioritize the surface over the content, just as we've been taught. We conceal our deviations and our weaknesses. We fix our gaze on the road ahead and we never, ever look away.

Scene 18. Levelling Out

Six months after Greta received her diagnosis, life has levelled out into something that resembles an everyday routine. It is 2015 and she has started at a new school. I've cleared my calendar and put work on the back burner.

Beata is in fourth grade. She lives and breathes music and dance. She is totally obsessed by the British girl-group Little Mix and her room is wallpapered with pictures of the four band members, Perrie, Jade, Jesy and Leigh-Anne. She's a little musical genius herself.

I can memorize an opera in two days if I have to, and almost no one I've ever met has a better pitch than I do – except for Beata.

She has sung in front of thousands of people, and been on a live broadcast at the Allsång sing-along at Skansen, and hit every note without a trace of nerves, despite millions of TV viewers.

I have never heard or seen anyone learn music faster than she does.

But while we're full up this year with getting back on our feet and taking care of Greta, Beata's having more and more of a tough time.

In school everything is ticking along.

But at home she falls apart, crashes. She can't stand being with us at all any more.

Everything Svante and I do upsets her. We guess it's because with us she can relax and drop the social game. She is highly sensitive and in our company can lose control and take out all her frustrations about sounds, tastes, clothes or basically anything that is too much and too difficult to take in.

Beata clearly is not feeling well. But it has still not occurred to us in what way. Neither do we understand how drained we are ourselves, simply from trying to get our days to fit together. Or the toll that level of fatigue takes on your judgement.

Scene 19. When War Moved In with Us

Autumn arrives and Europe is in the throes of the biggest refugee crisis since the Second World War. Although for the average person, it isn't much of a crisis, as long as you aren't working as an administrator at the Migration Agency or as a firefighter getting called out to extinguish blazes at Swedish refugee camps every other night.

Our family believes that no society in the world can manage the biggest refugee crisis since the Second World War without ordinary citizens rolling up their sleeves and trying to help. So we do what we can.

Beata and Greta want to do even more, so they propose that we let our summer house on Ingarö be used as refugee housing, and in November a small family moves in and we arrange bus passes and food, and we explain that they get to stay there until the asylum process is over. On the weekends we eat Syrian food together with all the neighbours and we look at pictures from Damascus.

Greta only sniffs at the food, leaning over saucepans and table settings. Beata sits on our loaned-out sofa, her back straight, with an exemplary smile. She bravely tastes her way through the Syrian cuisine. While Svante and I make our very best effort to be good guests.

But even if the war has moved in with us – even if it has remade our beds with donated Disney-print sheets from the Refugees Welcome transit housing in Sickla – it is still too far away for us to understand.And however we try, it still takes so much strength to take those baby steps forward that we are barely capable of processing anything else, however sincerely we might want to. We're just too tired.

Scene 20.
The Worst Bloody Mother in the World

'You bloody fucking bitch!'

Beata is standing in the living room, hurling DVDs from the bookshelf down the spiral staircase to the kitchen. There was a time when we had long, serious conversations about the meaning of such words, but that was a long time ago. Pippi Longstocking and the Teletubbies take the brunt of it. It's not the first time and it is definitely not the last.

'You only care about Greta. Never about me. I hate you, Mum. You are the worst bloody mother in the whole world, you bloody fucking bitch,' she screams as *Jasper the Penguin* hits me on the forehead.

Beata slams the door to her bedroom, kicks the wall a few times as hard as she can, and once again we are surprised at the incredible durability of double plasterboard.

We're fairly beat up too, but unfortunately we aren't as tough as the bedroom walls.

I'm not, in any case.

It's a whole lot harder to stay on your feet after the second blow. Now that our younger daughter is in need of extra care.

Greta's crash may have been more acute because she stopped eating, but this is a different kind of painful.

With Greta it was all kilograms, minutes, days, tables and structure. Everything was almost over-explicit and there was a measure of relief in what was square and ordered.

With Beata it's all chaos, compulsion, defiance and panic.

The only similarity is the time – as purely in terms of age, the explosion detonated at the exact same moment: pre-puberty. When the clock strikes ten or eleven years old.

Scene 21. Svante Solves Every Problem and Takes Beata to Italy

It only takes a few weeks for our everyday existence to once again be pulverized.

I've just started working at Stockholm's Civic Theatre and I fall fast. My reserves are drained and the adrenalin isn't there like it was with Greta.

Not at all.

'It'll work out,' Svante says, and decides to break the pattern and travel away with Beata so they can spend time together, have a little peace and quiet, whatever that is. And do all those things that people do when they're on vacation.

Greta can't travel because of her eating disorder and, besides, she now refuses to fly because of the climate thing.

'Flying is the absolute worst thing you can do,' she explains.

But she says that if it can help her little sister then of course they should go, and so Svante and Beata fly to Sardinia and take a rental car to a nice hotel close to the Strait of Bonifacio.

They jump in the pool and eat at a restaurant and Svante's rational thinking has come to the rescue yet again. The change of scene makes her happy and calm. It's working.

For a few hours.

Then she panics and wants to go home. There are lizards and sounds and it's too hot and she can't sleep.

'I want to go home now,' she cries.

'But we can't go home now. The flight leaves in a week.'

And that's a reality that she can't handle.

Beata gets a panic attack and she cries herself through the

night and the panic doesn't pass until breakfast. They try a little swimming in the pool, but all Beata does is cry and ask if they please can't go back home. She's scared and isn't doing well.

So Svante checks out of the hotel and packs up everything in just a few minutes and they jump in the car and drive the whole long way back to the airport with Little Mix blaring from the car stereo.

They make it in the nick of time for the afternoon flight to Rome and I book them on the SAS flight to Stockholm the next morning.

Svante finds a nice last-minute hotel close to the Piazza Venezia and from the rooftop terrace they watch the sun set behind St Peter's and it turns into a really nice picture on Facebook that gets a lot of likes and comments: 'Enjoy!'

Svante leaves yet another bit of his sweep-it-under-the-rug thinking in the Eternal City, and they fly home towards the glittering shores of Stockholm Arlanda. Beata is completely calm and content.

It is Midsummer Eve 2016 and all four of us walk home from the airport train shuttle with Moses on a leash. Greta and Beata each pick a bouquet of flowers along Kungsholms Strand: seven midsummer flowers – as is the old Swedish tradition – to put under your pillow so you will dream about your intended future love.

'You just released 2.7 tonnes of CO_2 flying there and back,' Greta says to Svante when no one else is listening. 'And that's the equivalent of the annual emissions of five people in Senegal.'

'I hear what you're saying,' Svante says, nodding. 'I'll try to stay on the ground from now on.'

Scene 22.
The Ballad of the Summer of 2016

It doesn't turn out to be a good summer. Neither of the children can go anywhere. Beata tried, and now she no longer wants to. We entice her with everything we can think of to do in the city, but she is not having it.

Everything we suggest is answered with a 'Shut up, you fucking idiot.' Meanwhile Greta can only eat a few things that have to be prepared in a special way in our kitchen. She can't eat around other people and even if her weight has increased and stabilized, she can't afford to miss any meals.

So we stay at home in the apartment. New experiences are now out of the question for Beata. She can't stand us and she can't stand hearing the sounds we make. Everything makes too much noise and holding all the thoughts in her head is impossible and there are too many thoughts and all the thoughts move too fast. Even Moses has to put up with her projections. He cowers under the piano and does his best to stay out of the way.

We just have to be real quiet.

Beata makes up her own games to play but the games are too hard. Instead they degenerate and take a compulsive turn, and when they don't work the way she wants them to work she rages at us because we are the only ones who are there. And when that's not enough the frustration grows. In the end, she develops compulsive behaviour around anything to do with sound, as a kind of defence mechanism.

The slightest little sound can cause an outburst. So the rest of us go out in the park or take little outings between meals.

We look at greenhouses and organic gardens and dip our feet in Lake Mälaren by Vinterviken bay.

Beata turns the day on its head, falling asleep at 5 a.m. and waking at 3 p.m.

A week or two goes by. Svante, Greta and I eat in the guest room on plastic plates so as not to make a sound. Everything rolls along. It's far from ideal, but it more or less works, and at least the days are passing by. We inch our way through the summer holidays from hell.

Then, one morning we wake up at seven because the whole building is shaking. Two neighbours have flown away on holiday and decided to have their bathrooms renovated at the same time.

They are refitting the sewage system in the concrete floor and the sound inside the building is deafening, but Beata can't go out and this is set to go on for more than two weeks.

In an instant the fragile foundation that we've built up collapses completely.

We beg and we plead. We curse and we swear.

The chairman of the housing association tries to help us, but obviously everyone has a full right to renovate their bathrooms, and everyone is doing what they can to limit the disruption. But that doesn't help us.

An unsustainable situation has become considerably more unsustainable and we take turns in losing our minds. Some days we try to set limits but that only makes everything worse, and somewhere in the midst of all this we get an appointment at BUP and I collapse inside the consultation room, hyperventilating.

Of course they want to help us but it's not easy during the summer holiday and instead we go to the Emergency Room at Sachsska Children's Hospital with scratched-up hands and arms and faces but they're closed. We carry on like that for several days, back and forth between BUP and the ER, and at last we get some tablets that will help Beata get to sleep in the evening.

But the whole family has already lost its footing.

I quit my job at the Civic Theatre and I'm prescribed antidepressants and sedatives, while waiting for summer holidays and bathroom renovations to end.

We scream. We kick down doors. We scratch. We pound walls. We wrestle. We cry. We ask for help and we somehow endure. But slowly, slowly an insight arrives, and with it, Beata's journey begins.

Scene 23. Between the Lines

The number of psychiatric diagnoses has increased exponentially in Sweden, as in many countries. The reason is, of course, that the number of complications has increased for those with various *suspected* diagnoses – complications that are very often stress-related.

More people simply have cause to investigate why their everyday lives don't function as other people's everyday lives seem to function.

More and more need tools to describe, for example, a functional impairment: and a diagnosis can be one of those tools. That's why diagnoses are a good thing. They save lives.

At the same time, the vast majority of us don't fully know how diagnoses function along with the fact that we are constantly being fed incorrect, stereotyped images which very often risk doing more harm than good.

Autism and ADHD and all the other neuropsychiatric functional impairments are not handicaps per se. In many cases, they can be a superpower, that out-of-the-box thinking you so often hear performers, artists and celebrities talk about. Performers like me, for instance.

But the complications that can arise because of a diagnosis can definitely be compared to a handicap; a handicap that is created by ignorance, incorrect treatment, discrimination or an inability to provide much-needed societal adaptation.

That handicap can be borne and carried by more then one person. As a family you can offer relief, plough the way, share the complications, and yes, that helps. With the right assistance and adaptation the problems often diminish as children get older. But if you don't get help, it is our experience that the problem often spreads quickly and whole families risk becoming more and more codependent or co-handicapped.

Tens of thousands of families in Sweden are experiencing this now – families that to a great extent are living parallel to Swedish society in a state of alienation that no one else seems to have the slightest clue about.

Of course, there are no economic interests represented in this community. There are no lobbying groups. Here, among the invisible children and the invisible families, hardly anyone can find the strength to talk at all. It requires too much energy – I almost can't bear to write about it myself. Because between the lines there is a story that can't be told. A story that absolutely no one wants to document, because those who have been there never want to turn back and be reminded about how it actually was. It's too tough. It's too ugly. It's too messy.

This story is way too humiliating for all involved – and that's why I have to tell it.

It's my duty because I have the opportunity to make myself heard. I have to speak out about the daily phone calls to teachers and parents to ease and facilitate everyday life. To the handicrafts teacher, the maths substitute, a classmate's father. Or about the thousands of late-evening emails to various school administrators sent to calm the children and help them sleep. Constantly forced to be the one who reminds others that what is best for the majority might be the worst for an individual.

P.E. classes that have to be changed, pop quizzes that aren't going to work. Outings that have to be cancelled. Medicine that runs out and pharmacies that haven't received the prescription. Every teaching assistant who never shows up, and sleepless nights, and phone queues at BUP, and notes from the school that never arrive.

Neighbours who complain, and broken walls, and disappointed friends who stop calling. All the phones, computers and Instagram accounts you want to throw to hell.

All the antidepressants and sedatives. Days when I'm not allowed to sleep at home because I make too many sounds. Concerts and music that I have to learn in the cellar storage room. Friends making comments on Snapchat.

All the days when you don't have the energy, or desire for anything. All the days that are a single, hopeless darkness. All the days – every day for five years – that our family could not eat together, or were barely able to be in the same room.

The stickers from the neo-Nazi Nordic Resistance Movement in our building's foyer, aimed at my family because I've chosen to take a public stand against racism. The pictures of our building posted on the internet to threaten us, and the evenings we went into the BUP emergency clinic and were given a cheese sandwich in the waiting room only to go home again because it was all just an attempt by us to set limits.

All the stay-at-homes: children who no longer manage to go to school.

Someone has to talk openly about a school system where one in four pupils ends up excluded. Pupils who end up without a diploma because special-education private schools that make millions in profit 'failed' to hire any teachers. The world's most profitable failure.

Children with autism who are forced to go to a school where 82 per cent of those like them are bullied. All the crisis meetings with schools and all the parents and teachers who get burned out. Everyone who is worse off than we are.

The connections between autism and depression, and children who take their own lives.

The darkest statistic: girls with anorexia.

And all the bottomless sorrow, sorrow, sorrow over all these childhoods, lost because we live in a society that excludes more and more people with every day that passes.

A society to which we devote all our time and energy adapting ourselves, and no, it doesn't require a neuropsychiatric functional difference to be able to see what is sick around us.

But the fact remains, sometimes the healthiest thing a person can do is to fall apart. The problem is that it doesn't help in the long run.

So we don't give up.

Whatever happens, we will never give up.

We try.

We heal each other.

Perhaps we will never be fine but we can always get a little bit better, and there is strength in that. There is hope in that.

Scene 24.
Street Dance

It is Tuesday and we're just starting to recover after yet another abysmal weekend.

Last Friday a new teacher came up to Beata and asked why she looked so tired, and when she usually went to sleep at night.

'Midnight,' Beata answered. The teacher hit the roof and gave her a long lecture – with the best intentions, of course – about the proper bedtime and how many hours you need to sleep to cope with school, and everything she said stressed Beata out so much that she couldn't sleep all weekend. It took three days for the family to fall apart again like a house of cards.

But today it's Tuesday and it's time for dance class.

It's important to be there on time, as the stress of arriving late can be so severe that it's impossible to get going at all.

So we give ourselves plenty of time.

Beata has to avoid all the cobblestones, you see.

She has to lead with her left foot and if she gets it wrong she has to start over. And I have to walk exactly the same way, which is hard because I have longer legs, and we're probably quite the funny sight, walking the street back and forth. Back and forth.

It's no more than a kilometre by foot but it takes almost an hour and it's only with me that she has these compulsions. And I understand. Because I was exactly the same with my mother – all of my tics got much worse in her company.

When we arrive we discover that there's a substitute teacher today, which is not good, because it will be different, and Beata doesn't like different. I sit down on the floor outside and wait.

Every Tuesday I sit there for two hours. I can't move from the spot, not even to go to the toilet, because Beata will worry. She has to be able to see me through the crack in the door at all times.

I feel the bass vibrate in the walls and the floor. Immediately there's a stab of worry in my gut. The volume is not usually this high. I answer emails on my phone, try to do something useful. After a while I sneak in and peek through an opening in the drapery inside the door. The music is thundering in there and eight girls are doing street dance while the substitute stands at the front shouting out dance steps. The ninth girl isn't dancing at all – she's standing in the middle of the room holding her ears and sobbing. Her whole body is shaking.

I rush in and ask them for the umpteenth time to lower the damned volume and don't you see that she's crying?! But the substitute doesn't understand why he should care, so I take Beata with me and go home. And the weekly street dance becomes yet another in a long series of failed group activities.

But before we leave I get to hug her.

For a long time. She is crying desperately and it's awful, but at least I get to feel like a mother who is needed by her child.

It's the first time in ages that I get to hold my beloved little one in my arms. It's like coming home after a life in exile.

It's the best moment.

Of all.

Scene 25. The Low-arousal Approach

It's autumn when Beata undergoes an evaluation for various neurodevelopmental disorders. We are having our last individual BUP meeting before the assessment with Beata and the school staff.

'I remember one incident when it was time to vaccinate Beata at school and how she obsessed about this for weeks,' Svante says. 'Sometimes she would start crying uncontrollably because she was going to get that needle. So when the day finally came I went with her to school and followed her to the nurse and during that whole time we were in there, she didn't show even the least bit of fear or discomfort. In fact she showed no expressions at all. She took off her sweater, extended her arm and took the syringe without batting an eye. She looked like she was watching a really boring movie on TV. She got her plaster, put on her sweater and went back to class as if what had happened was the most obvious and meaningless thing in the world. But when she came home in the afternoon she fell apart and had a long, angry outburst.' Svante stammers a little as he tells this story, as he does when he's upset.

Several different diagnoses are partly applicable, but nowhere does she fit with the criteria that are required to get a diagnosis.

'You can have 90 per cent ADHD, 60 per cent autism, 50 per cent oppositional-defiant disorder and 70 per cent obsessive compulsive disorder,' the psychologist explains. 'So together it adds up to over 100 per cent neurodevelopmental disorder but it's still no real diagnosis.'

When she's done talking I discover that for the first time in fifteen years Svante is crying in public. He doesn't cry often, but now he can't stop.

'You have to help her,' he says, crying and sobbing. Again and again and again.

At last Beata is diagnosed with ADHD, with elements of Asperger's, OCD and ODD.

If she hadn't been given a diagnosis we, together with the school, would not have been able to make the necessary adaptations to help her function and feel okay again. If she hadn't been given a diagnosis I wouldn't have been able to explain to all her schoolmates' parents and all the teachers and adults. If she hadn't been given a diagnosis I wouldn't have been able to keep working. If Beata hadn't been given a diagnosis we would never have been able to write this book.

That's how crass reality is. The difference is like night and day.

But now she has it and for her it is a fresh start, an explanation, a redress, a remedy.

Our daughter goes to a good school. A school with resources, knowledge and competent personnel. One of the very few that have started to take inclusion, functional differences and individual adaptation seriously. But it is still the individual, voluntary efforts by teachers that make the real difference. She has marvellous teachers who make everything work. She doesn't have to do homework. We drop all activities. We avoid anything that may be stressful.

And it works. At home we learn that a low-arousal approach works best. Whatever happens we must never meet anger with anger, because that, pretty much always, does more harm than good. We adapt and we plan, with rigorous routines and rituals. Hour by hour.

We try to find habits that work. The unforeseen can of course lead to everything collapsing again, but if it does, then we start again from the top. We divide up. Each of us takes a child. We live in different places.

All families have a hero. Beata is ours. When Greta was feeling her worst it was Beata who had to take a few steps backwards and manage on her own. If she hadn't done that we wouldn't have made it. Without her, it wouldn't have been possible.

I'm the one who is closest to her, because I'm her mother. And we are so uncannily alike. I'm the one who understands her best, and she knows that. Although she'd never, ever admit it.

Sometimes I make mistakes. Sometimes I'm the child. I can't always cope with a low-arousal approach.

But I try.

And I love her to the end of time and beyond.

Scene 26. Higher Ground

The fact that our children finally got help was due to a great many factors.

In part it was about existing care, proven methods, advice and medication. But it was primarily thanks to our own toil, patience, time and luck, that Greta and Beata found their way back on their feet. Along with the fact that a number of individuals broke the rules and did things they weren't supposed to do because they knew it was the right thing to do.

And we can't continue like that.

A functioning society cannot rely on luck or civil disobedience. Most parents don't have 250,000 followers on social media like I do . . . Most parents can't be at home full-time without going on sick leave. Most parents – needless to say – don't have the privilege and advantage of status and celebrity.

II

Burned-out People on a Burned-out Planet

I can't take any more.
Or yes, of course I can, but you get the general idea.

– Nina Hemmingsson

Scene 27. Denial

The scent of fabric softener wafts from a ventilation grate on Fleminggatan.

Stockholm. January.

Piles of Christmas-tree graveyards on every corner.

A never-ending cold rain accompanies me along grey, slushy streets. During the Christmas holiday season the city is almost empty; everyone who lives here is somewhere else.

Everyone is in Los Angeles or Thailand. In Florida or Sydney. On the Canary Islands or in Egypt.

We Swedes are incredible. We stand up for almost everything that can be stood up for. We fight for refugees and vulnerable people and against injustice everywhere.

From an ecological perspective, however, we are very far from incredible – and people like me are among the worst there is.

'You celebrities are basically to the environment what anti-immigrant politicians are to multicultural society,' Greta says one morning at the breakfast table.

It's not a nice thing to say to someone who really believes in cultural diversity. But I guess it's true. Not just of celebrities, but of the vast majority of people. Everyone wants to be successful, and nothing conveys success and prosperity better than luxury, abundance and travel, travel, travel.

'At the same time, if I get sick or unpopular I won't get a penny,' I say in my defence. 'Moral responsibility can't always be demanded from you just because you have certain opportunities to make yourself heard and set a good example.'

But Greta doesn't agree. She scrolls through my Instagram feed. She's angry.

'Name one single celebrity who's standing up for the climate! Name one single celebrity who is prepared to sacrifice the luxury of flying around the world!'

'They fight for other things,' I say, without coming up with any useful arguments whatsoever.

'Okay! Name a single thing that they are fighting for – except possibly against full-scale nuclear war – that we wouldn't be able to fix later on in the future. If we really wanted to.'

She's right, of course. If we destroy the climate we'll never be able to repair it, and soon future generations – or present ones for that matter – won't be able to make everything right, and undo our mistakes, however much they might want to.

And sure, we all fight for the wrong things. Or rather: we fight for the right things but as long as our lifestyle is at odds with the most important issue of all, there is a great risk that in the end our struggle will have been in vain. For nothing.

Surely, not everyone needs to become a climate activist. But at a bare minimum we could all stop actively destroying our environment and planet, and stop showing off that self-same climate destruction as trophies on social media.

I am a big part of the problem myself.

Not three years have passed since I was posting sun-drenched selfies from Japan. One 'Good morning from Tokyo' and tens of thousands of 'likes' rolled in to my brand-new iPhone.

On the flight home I spent a whole day staring out over Siberia and the Arctic Ocean while the jet engines sang the monotonous song that plays a unique minor role in the liberation of the tundra's greenhouse gases from their 100,000-year permafrost sleep. The awakening of the mighty methane dragon.

Something started to ache inside of me. Something I'd previously called travel anxiety or fear of flying but which was now taking on another, clearer form. Something was wrong.

But I had performed for 8,000 people and the concerts had been recorded for Japanese TV, so my trip had served a purpose, I convinced myself.

As if our biosphere and ecosystems care about Japanese TV.

Denial is a mighty force.

Scene 28. Gluttony

From a human point of view, a balanced, functioning atmosphere is a finite asset; a limited natural resource that belongs equally to all living things and beings. At the current emissions rate this natural resource will be used up in eight to sixteen years time, depending on if you aim for the 1.5°C or the 'well below 2 degree target' set in the Paris Agreement. And how many per cent chances do you wish to give us?

But it is much worse than that.

To maintain a safe and sustainable climate the concentration of CO_2 should not exceed 350 parts per million, according to leading researchers. At the present time we have already passed 415 ppm and within ten to twelve years we're expected to reach 440 ppm.

And so on.

The emissions for a person who flies economy class, Stockholm–Tokyo return, are 5.14 tonnes of CO_2, according to Arlanda Airport's own carbon-offsetting programme.

The combined flight time of about twenty-five hours corresponds – roughly calculated – to an individual's consumption of 200 kg of beef.

Our new habits have undeniably shone a new light on the gluttony of both the ancient Romans and eighteenth-century French aristocrats.

The average emissions for an inhabitant of India, according to the World Bank, are 1.7 tonnes per year. In Bangladesh the figure is 0.5 tonnes.

And no, soon we will no longer be able to talk about solidarity and equality without weighing up our own ecological footprint. To stand up for justice is a mandate that is in the process of slipping out of our hands.

Scene 29.
Symbiosis

I should have written a cookbook instead. A book about cookies and my favourite composers. Or a real autobiography. A singer's memoirs.

Nothing about burn-out, medication or diagnoses.

An agreeable book. Maybe about yoga. In which, naturally, I would tackle some agreeable environmental issues as well: plastic bags, food waste or something else that doesn't come across as too uncomfortable or awkward for anyone.

A positive book that in no way touches on things like eating disorders, environmental disaster or depression. Or how on certain days you can't get out of bed because you can't, don't want to or don't have the energy. How on certain days you have thoughts you shouldn't be having.

I should not have written a book about how I felt.

I should not have written a book about how my family has felt for long periods during the past few years.

But I had to. We had to. Because we felt like shit. I felt like shit. Svante felt like shit. The children felt like shit. The planet felt like shit. Even the dog felt like shit.

And we had to write about it.

Together.

Because once we realized why we were feeling the way we did, we started to feel better.

We had to write about this because we are among those who got help. We got lucky and sometimes I think that we are going to come out of this strengthened. Strengthened and whole. I believe that fairly often.

And it's time we all started talking about how we're really doing. We have to start telling it like it is.

We live in a time of historic abundance. The world's combined resources have never been greater. Just like the chasms dividing rich and poor. Some have so insanely much more than they need. Others have nothing.

At the same time the world around us is only faring worse. The ice is melting. Insects are dying. The forests are disappearing and the oceans and other ecosystems are struggling more and more each day.

Like so many people around us.

People who have fallen apart like we fell apart, people who are still tattered and torn. Our friends.

The ones who couldn't keep up the pace.

Those who didn't fit the mould.

Those who didn't have the good fortune to meet the right doctor.

Those who didn't show up in the statistics.

All those countless people who truly live in symbiosis with the planet they inhabit. Not the symbiosis we usually talk about, the one we associate with a down-to-earth life in harmony with nature.

This is about a new unanimity: a new chord. This is about burned-out people on a burned-out planet.

And such stories don't belong in cookbooks.

Scene 30. Astrophysics

It takes 23 hours, 56 minutes and 4.091 seconds for the earth to revolve on its axis and complete a full day. Sometimes it feels like it's going a little bit faster, although the velocity is always exactly the same, down to the millisecond. On the other hand there are things that spin around us and definitely do increase in velocity – our lives, for example.

When I was little they used to say that a computer could never replace a human.

'Just look at chess! A computer can't beat a person,' people said.

Then, in 1990, a man by the name of Ray Kurzweil maintained that because the world's computing capacity was doubling every year, a computer would beat the world's best chess player before 1998; this was a matter of pure logic.

And sure enough, in New York on 3 May 1997, in one of the most famous chess games in history, the reigning world champion, Garry Kasparov, was defeated by IBM's software program Deep Blue.

So that was that.

Today Ray Kurzweil is Director of Engineering at Google and makes assertions such as a child equipped with a smartphone in the African countryside has access to more information than the American president had only twenty years ago. Computers being in possession of human-level intelligence is, according to Kurzweil, also only a matter of time; a mathematical truism that will materialize no later than 2029.

This says something about the speed at which our society is changing right now.

But it doesn't say everything.

We experience more, we feel more, we have more and stronger opinions. On social media we debate social issues at a speed and to an extent that makes the 1990s seem like the Middle Ages.

Nothing gets to just be, everything has to be taken to extremes. Everything has to be brought to a head.

We produce more. We consume more. In fact, whatever it is we do, we do it more. Much more.

Scene 31. Think Big and Kick Ass: In Business and Life

That's the title of one of the current American president's best-selling books.

Think BIG!

Donald Trump embodies some of the worst parts of our society. He is the end of the road for our time, but what he represents is of course nothing new. We've all been living in his world for a very long time. In the winner's world. A world where everything must expand.

The world is like a carousel spinning at increasing speed – faster and faster.

But how fast is fast enough? Are we ever going to reach a critical point, a point where we can no longer close our eyes to all those who aren't able to adapt to the speed; the ones who get flung off in the middle of the ride? All those we sacrifice in favour of a society with endless economic growth? Which, to be fair, leads to a higher standard of living because the driving force to be a little bit better off and get a little closer to the dizzying upper echelons drives everything forward. And sure, that all might sound pretty reasonable sometimes, if you just close your eyes to the fact that we are well on our way to sawing off the branch we are all living on.

For the fact is that, in the midst of all the upward-trending growth curves, a great many people are feeling worse and worse. Involuntary solitude has become a chronic public-health issue. Burn-out and mental illness are no longer a global ticking health bomb – the bomb has already exploded.

Scene 32.
Stress Disorders and Health Statistics

In Sweden, mental-health issues in children aged 10 to 17 have increased by over 100 per cent in 10 years. According to the Swedish National Board of Health and Welfare's report of December 2017, almost 190,000 children and young adults in Sweden suffer from some form of mental illness. Girls and young women fare the worst, with 16 per cent having had some form of contact with adolescent psychiatry services. That's almost one in six girls in Sweden.

The number of ADHD and autism diagnoses have each more than doubled in the past five years.

There is talk of tens of thousands of stay-at-home children. The grey zone, where these children reside, is enormous and the number of unreported cases is huge – because no one wants to be here, no one wants these figures to exist since they speak of fundamental failure for everyone involved. But in the incomplete statistics we sense the contours of a catastrophe.

And nowhere are there any signs that the trend is in decline. Or levelling out.

It's accelerating.

Scene 33. In a Skirt and Boxing Gloves

We try to find out why this is the state of things. We search and think that nowhere on earth are women equal to men and that the evidence of this is everywhere, all the time. Certain examples are more obvious than others, but once you start looking there seems to be no end to the discoveries.

To generalize, the strongest woman cannot equal the strongest man in terms of muscle mass or lung capacity. But that's not to say women aren't stronger than men.

It's all about how we perceive strength and which qualities we value most. But no one can deny that the qualities we traditionally associate with success and happiness are strongly linked with male physiology.

Higher, bigger, faster, stronger. More.

And when we say that we want an equal society, what this often means in practice is that, if women are going to succeed, we must embrace those male qualities that in reality we are never going to be able to fully appropriate. We have to become like men.

We have to compete against men on men's terms. Just like the iconic image of the woman who has rolled up her sleeves and is flexing her biceps, we all have to become women in skirts and boxing gloves: symbols for a struggle that in the end we can never overcome. And if, against most expectations, we do so anyway, we're probably going to be perceived as unwomanly – too strong, too educated – and whatever we do will almost always be wrong in one way or another. And society will no doubt continue in exactly this way until we start to seriously expose the structures that rule this world.

The structures that for every passing second fling more and more people off this merry-go-round. The structures that often expose us to real physical danger because we are forced to become something or someone that we are not in order to be classified as successful. Or to simply fit in.

According to the Swedish Social Insurance Agency, the number of people on medical leave with chronic fatigue has doubled six times since 2010. Over 80 per cent are women.

These figures speak for themselves. It's a dramatic language.

And the fact that they don't take up more space in public debate and the media also tells a story. A story that – once again – explains to future generations what and who are prioritized.

And what and who aren't.

Feminism, on the other hand, means many different things to many different people, and many are taken aback when it is mentioned in a sustainability or environmental context. The connection deserves a separate section in every well-stocked library, but here it is probably enough to say that women and highly sensitive people are gravely over-represented in those bleak statistics that constitute the downside of our competitive society.

Scene 34.
A Historic Transition

As we've said, we find ourselves in a sustainability crisis. But we also find ourselves in an acute climate crisis. There is almost no one who still denies this crisis, which is a good thing. However, the problem is that it's a big step from acknowledging the existence of the crisis to understanding what it actually means. A very big step.

We as human beings are currently in the middle of that step – in a void where everything can continue as usual.

We think we know what the crisis means.

Everyone assumes that everybody knows.

Scene 35. A Letter to Everyone Who Has a Chance to Be Heard

My name is Greta and I am fifteen years old. My little sister, Beata, will turn thirteen this autumn. We can't vote in the parliamentary election even though the political issues now at stake are going to affect our whole lives in a way that can't be compared with previous generations.

If we live to be a hundred then we're going to be here well into the next century, and that sounds really strange, I know. Because when you talk about *the future* today, it usually means in just a few years' time. Everything beyond the year 2050 is so distant that it doesn't even exist in our imaginations. But by then my little sister and I – hopefully – will not even have lived half our lives. My grandfather is ninety-three and his father lived to be ninety-nine, so it's not an impossibility that we're going to live long lives, too.

In the years 2078 and 2080 we will celebrate our seventy-fifth birthdays. If we have children and grandchildren, perhaps they'll celebrate those birthdays with us. Perhaps we'll tell them what it was like when we were children. Perhaps we'll tell them about all of you.

Perhaps they are going to wonder why you, who had the chance to be heard, didn't speak up. But it doesn't have to be that way. We could all start acting as if we were in the middle of the crisis we are in fact in.

You keep saying that the children are our future, and that you would do anything for them. Such things sound full of hope. If you mean what you say, then please listen to us – we don't want your pep talks. We don't want your presents, your package holidays, your hobbies or your unlimited options. We want you to seriously get involved in the acute sustainability crisis going on all around you. And we want you to start speaking up and telling it like it is.

Scene 36.
The Luxury Trap

In Sweden, according to the Swedish Environmental Protection Agency, we release approximately 11 tonnes of CO_2 per person per year if we take in to account both what we emit here at home and what we consume abroad. These are known as our consumption-based emissions. According to the WWF's latest *Living Planet Report*, our ecological footprint is among the largest in the world, and if everyone were to live like us it would require 4.2 Planet Earths.

We tell ourselves that we can still choose, that our emissions can be offset somewhere else. Or that we can make our own partial deals with nature – such as becoming a vegan in order to keep flying, or buying an electric car to keep shopping and eating meat, or paying climate compensation for things that we plan to do tomorrow when, from a sustainability perspective, we've already so far in debt that the numbers run far beyond our comprehension.

The hard truth is that our ecological credit ran out when we passed 350 parts per million of CO_2 in the atmosphere. More precisely, in 1987.

Scene 37. Windfall Apples and Radioactive Waste

'Sweden will become the world's first fossil-free welfare state,' the prime minister said in the government declaration of autumn 2017. It sounded great. Almost as great as the government declaration two years earlier, when he basically said exactly the same thing. But the truth of the matter is that in those two years absolutely nothing had been done to move us any closer to that green, fossil-free welfare nation.

According to a 2018 study by the Swedish Society for Nature Conservation, the Swedish environmental budget amounts to 11 billion Swedish kronor ($1.2 billion). At the same time, the government budget contains environmentally damaging subsidies of 30 billion kronor – subsidies that make it cheaper to emit greenhouse gases. So, basically we send out a fire engine full of water at the same time as we send out three tankers full of gasoline – both meant to extinguish a massive fire.

But.

'Fossil-free' is undeniably an extremely good word. Both growth-friendly and radical. And at least as powerful as 'sustainable', although with considerably fewer requirements. The fact is, 'fossil-free' can mean anything from solar energy and organic windfall fruit to deforestation, carbon trading and radioactive waste.

Investing in words like 'fossil-free' means that we can push words like 'change' into the distant future and instead delay our ecological-payment due date for many more years to come. That way we can carry on like before while making bold, environmental statements like 'we are among the most sustainable societies in the world.'

Scene 38.
The Fine Print

We often hear that soon we're going to have to get by on a measly 2 tonnes of CO_2 per person per year. Or that here in Sweden we'll be forced to reduce our emissions to a tenth of what they are now if we're going to comply with the Paris Agreement. But all such calculations depend on things far beyond anyone's control. Like inventions that have not yet been invented. Such as gigantic machines that could suck our carbon dioxide out of the atmosphere or a sustainable forest and agriculture industry that doesn't exist today.

Or that none of the earth's other 7.7 billion people will somehow come up with the idea of raising their standard of living and start mimicking the daily habits we now consider to be our right.

'The two tonnes figure is unfortunate,' Kevin Anderson at Uppsala University says. 'It actually means nothing and I think it's an old leftover from one of the first UN IPCC reports, where they took simplistic emissions figures and said that they needed to be cut in half. Our CO_2 emissions must go down to zero, that's the bitter truth.'

So the 'nightmare' figure of 2 tonnes may very well not be enough.

On the other hand this doesn't mean that everything is hopeless or too late. It means we have to change our way of life. Starting right now.

Scene 39.
A Dream Play

The surreal unreality is almost the worst part. There are moments when you wonder if you've lost your mind.

Gone crazy.

All the countless occasions when you realize that what makes up our everyday lives – everything that we call normal – is often as far from normal as you can imagine.

All the incomprehensible moments when everything around you is transformed into a theatre or movie production.

Like an air-conditioned hotel room in a boiling far away major city. A shopping centre with 400 stores. Driving through a snowstorm until you reach shelter in something we call a motorway tunnel. A supermarket with groceries from every corner of the world. Or the relief in meeting the friendly gaze of a Swedish-speaking flight attendant who nods in recognition as you board an SAS flight somewhere on the other side of the globe.

All the little things that we – and all those who populate our everyday lives – perceive as natural, keeping us safe and out of danger.

Now most of all this resembles a scenic design; magnificent decor from the age of humans: the Anthropocene.

The party is over.

The play is over.

And so a window swings open and a new light fills the room. Unreality becomes reality.

Fabrics and curtains flutter in the wind. Props whirl around in drifts. Masks and dramatic interpretations are done away with and the reality rings out over stage and auditorium. Everything is turned upside down. Back and forth.

We who have separated our culture from nature and always placed the facade and surface in the centrestage spotlight have suddenly crossed an invisible line. One by one, we are slowly getting off the stage, and yet are still there, as if standing in the middle of a play as it frantically continues around us.

But this performance is already over and now it's time to start changing our behaviour. Tear down the fourth wall. Stop pretending.

A society that prioritizes the surface over content can never be sustainable, and we're not going to be able to solve the climate and sustainability crisis if we don't confront the culture that forbids us from talking about how we're really doing – all the things we have chosen to sweep under the rug for decades and centuries.

Scene 40. The Art of Lying

'Sometimes we human beings mess things up.' The prime minister is talking about the climate in a direct broadcast from Parliament.

'He's lying,' Greta says, getting up from the sofa in front of the TV. 'He's lying!'

'What do you mean?' I ask.

'He says that it's we people who have messed it up, but that's not correct. I'm a person and I haven't messed it up. Beata hasn't messed it up, and you and Dad haven't done it either. At least not that much.'

'No, I guess you're right about that.'

'He's just saying that so we will carry on as usual, because if everyone is guilty no one is guilty. But someone *is* actually guilty, so what he's saying isn't true. There are like a hundred big companies that account for almost all emissions. And there are a few extremely rich men who have made thousands of billions destroying the whole planet, even though they know all about the risks. So the prime minister is lying, just like all the rest.'

Greta sighs.

'It's not everyone who has messed it up. There are a few, and to be able to save the planet we have to take up the fight against them and their companies and their money and hold them accountable.'

Scene 41. Green Growth

Whenever we hear a politician or a sustainability manager talk about the climate or the environment, they're saying the same thing – namely that our emissions have to be reduced.

And they *must* be reduced. By 10–15 per cent per year for a country like Sweden, if we're going to achieve the Paris Agreement's 2°C target, while taking into account the aspect of global equity.

The problem is that those emissions that 'have to be reduced' – with some isolated exceptions – never are. And because these isolated exceptions only happened as a result of a few global financial crises, then it's not so strange that anyone who thinks in a slightly shorter time frame than a museum of natural history, never felt that those emissions were worth striving for. Which basically means everyone here on planet earth.

So emissions continue to increase, even though we've been at levels way over those required to maintain a stable climate for a very long time. The last time we had this quantity of CO_2 in the atmosphere the ocean sea level was about twenty metres higher than it is today.

And no, of course it's not by chance or coincidence that emissions are increasing. It's a conscious decision and this will continue until we decide that our only overall goal is no longer economic growth, but instead radically reduced emissions: shutting down the oil rigs as fast as possible, keeping the coal and carbon in the ground and adapting ourselves to the very same new reality that the world's research community has been telling us about for three full decades.

This is not to say that green, sustainable growth is not desirable, possible or welcome.

But right now reduced emissions must be our primary goal, because all our margins have been used up.

Scene 42. The Really Boring Part

Svante is sitting by the computer, rubbing his face with his two hands. A rough draft of the book went out for review and now we're assessing the results. He turns towards Greta.

'Okay, there are some who say that the reading gets a bit heavy around Scene 41; they think it's a lot more fun when you and Beata are present. Can something be added there?'

'Like what?' Greta asks. She has just selected a few pictures of pigs in slaughterhouses that she hopes can be included in the book along with a discussion about the billions of animals that live out their short lives on a conveyor belt, because we humans have assumed the right to industrialize life itself.

'Can we write something about you instead?'

'No,' she replies curtly. 'Lots of personal and other stuff comes later. Mum's burn-out and all the things that people love to read about celebrities. This is a book about the climate and it's supposed to be boring. I don't care. The readers will just have to put up with it.'

Scene 43.
Business as Usual

An information war is being waged about the future conditions for survival in major parts of Planet Earth. Researchers and environmental organizations say one thing; the lobbyists and most of the business community say another.

Thanks to the media's lack of interest, our future ecological state has been reduced to a political game where it's word against word, and the most popular wins. And guess which climate and sustainability story sells the best? The one that demands changes or the one that says we can continue shopping and flying for all eternity?

Guess which story most politicians get behind? The problem is that when it comes to the more popular option, they tend to leave out one or two little details. Such as that the crisis is foremost a crisis and not just an opportunity for new financial progress. So the greatest threat that humanity has ever faced ends up drowning in a sea of empty promises about future eternal 'green' economic growth.

Here, the roaring sound from melting ice-caps and glaciers remains mute. The stories about how global industrial agriculture risks our future remain untold. And here there is no one talking about how the world's rainforests are being cut down so ruthlessly that they are now feared to release more CO_2 than they absorb.

One of humanity's most winning characteristics is our capacity to adapt to change. And even if that change is not always welcome, we almost always get to grips with it once we're faced with critical events.

The planet's sixth mass extinction, the one that has begun all around us, is a crisis. The melting process in Greenland, the Arctic and Antarctic is a crisis. The fact that we have lived through a time of extremely unusual climatological stability – a stability that enabled the rise of civilization – and that our lifestyles mean that era is now behind us is a crisis. But those stories go unheard because we drown them in a tidal wave of trash.

A new world order is at the door. Astronomical economic interests are at stake: lies, half-truths and creative statistics are shared diligently from every conceivable direction. Emissions are set against emissions, even though *all* emissions must be reduced dramatically.

Aviation blames cars. Agriculture blames aviation. Motorists blame shipping. Because it's easier to blame others than to take stock of your own house. And it's like there's always someone else who should be doing more than us. There's always some international legislation or some little detail that we ought to be focusing on instead of taking action. Our very future is at stake and the best we can come up with is a childish 'Yeah, but what about them! True, our emissions won't be reduced, but all that business as usual really benefits everything and everyone!'

Everything except future life on earth, of course. But who cares about that?

We have placed our fate in the hands of good intentions. And we have done it in a time when even hospital patients and schoolchildren with functional disorders are supposed to contribute to financial profitability.

What could possibly go wrong?

Scene 44. Lip Service

'At least Donald Trump is honest. He prioritizes new jobs and more money and blows off the Paris Agreement so everyone calls him an extremist. And rightly so, but we're all doing the exact same thing,' Greta says.

We're watching the party leaders' debate on Swedish television.

Svante goes out with the dogs. He can't bear to watch, gets too angry.

'Our emissions are among the highest in the world,' Greta continues, upset. 'And now almost all the party leaders are standing there saying that we mustn't focus on our *own* emissions but instead help neighbouring countries who are evidently worse than we are. Even though our ecological footprint is so much higher than theirs! And no one's saying anything?!'

She's sitting on the couch with the computer on her lap. Outside the window the summer heat is already here, even though the calendar page has only just been turned to the month of May.

'We've got one of the biggest ecological footprints in the would per capita,' she continues. 'Should we help others? The USA and Saudi Arabia then? We're the ones who need help. And the programme hosts aren't saying anything, because they probably don't know that we've exported our emissions to other countries. No one knows, because no one talks about it. Everyone complains about Trump's alternative facts but we're probably even worse than he is, because we fool ourselves into believing that we're doing good things for the environment.'

The next day the newspapers do a fact check of what was said during the debate. But what is checked is quite different from what we talked about – namely the rate at which the ice is actually melting. Is it really 200,000 m^2 of ice that is melting *every month* or is it possibly a little less? No one is disturbed by the fact that most of the party leaders understate Sweden's emissions by more than half. Greta reads the article at the breakfast table and comments: 'One day we miss our emissions targets by miles. The next we're going to expand all our airports, triple the number of passengers and build *climate-smart* highways. They say that climate-change deniers are idiots. But everyone is a climate-change denier. Every single one of us.'

Scene 45. The Optimists

In the summer of 2017, six leading scientists and decision-makers on environmental issues wrote in the journal *Nature* that humanity had three years left to get the emissions curve to change direction – to take a steep downward turn. *Three years to safeguard the planet* and if we are not going to achieve that, then we're at great risk of missing the Paris Agreement's well below 2°C target and thereby placing ourselves in great danger of starting a negative spiral of climate disasters far beyond our control.

That is, as long as the world is not prepared, by the year 2025, to close almost all factories and let all cars and airplanes stay parked and slowly rust, while we eat what's left in the pantry. And the authors of this article aren't usually counted among the alarmists.

'These are the optimists,' the *Washington Post* wrote.

Now more than a year has passed and there are almost no signs of the revolutionary changes required and the leadership we so desperately need. *Sweden is a pioneering country*, we often hear. But the truth is that there are no pioneering countries. At least, not in our part of the world. Because our climate and environmental struggle has got nothing to do with rescuing the climate – we're fighting for the possibility to keep on living the way we do.

Scene 46.
Anno Domini 2017

In the year 2017, 9 million people died from environmental pollution.

Over 20,000 researchers and scientists issued a sharp warning to humanity and explained that we're heading for a climate and sustainability catastrophe; time is running out.

In the year 2017, German researchers determined that 75–80 per cent of insects had disappeared. Not much later came the report that the bird population in France has 'collapsed', and that certain bird species have been reduced by up to 70 per cent because they have no insects to eat.

In the year 2017, forty-two individuals had more money than half the world's population combined and 82 per cent of the world's total increase in wealth went to the richest 1 per cent.

Sea ice and glaciers were melting at a record rate.

65 million people were displaced.

Hurricanes and torrential rain claimed thousands of victims, drowned cities and smashed whole nations to bits.

It was also the year when the emissions curve again turned upwards, at the same time as the quantity of CO_2 in the atmosphere increased at a velocity which, from a larger geologic perspective, can only be compared to pressing the warp button in a *Star Trek* movie.

Scene 47. No, Please, No More Pieces about the Climate

'The climate is a burning issue. It's super-important. But I want you to write about other topics.'

Once a month I write for *MittMedia* and *Dalarnas Tidningar* and today is the November deadline. My very shrewd and intelligent editor has just received another 750 words about the climate and between the lines she is venting her frustration.

'I don't want any more pieces about the climate!'

I could not agree more. I don't want any more pieces about the climate either. I want to write about other things. Things that the newspaper and I have agreed to focus on. Culture. Thriving rural towns. Humanitarian projects. Anti-racism. The municipal music school, or really anything at all.

I want to be like other columnists who write about everything imaginable, and who might go against the current every other month with a column about climate change, only to return to writing about hospital food, Muslim calls to prayer in the city of Sundsvall or whichever social phenomenon is on everyone's lips right there and then.

I want to think in line with everyone else when they list the most important issues before the election campaign and mention five or ten different matters that don't get enough attention and that we ought to be talking about more. I too want to list the climate threat as number three, perhaps after the school system and healthcare.

But I can't because now things are the way they are. And however much I try to ignore it, it just doesn't work. I'm fascinated by those who manage to engage with other issues. It's a little like discovering people in the early twentieth century passionate about issues that don't in some way concern the universal right to vote, the condition of the working class, women's liberation or the right to belong to a trade union.

Except that this is much, much more dramatic. Because a hundred years ago you didn't have a gigantic clock counting down the fate of all future generations right in front of your eyes.

'The issue is too big,' Svante and I often hear. 'It's impossible to take in.'

And this is both true and not true at the same time.

It's actually fairly easy to get involved in the problem if you're so inclined. If you're prepared to make the sacrifice, to renounce some privileges and take a couple of steps backwards.

Because the climate issue in itself is not all that difficult or hard to fathom. But it is too uncomfortable.

A bit like being cozily curled up, deep asleep in a warm corner of a big sleeping bag inside a rain-drenched tent. You don't really want to get up and deal with the problem. You want to go on sleeping. Like everyone else.

My last column in *Dalarnas Tidningar* is about the fact that *MittMedia* continues to publish unrefuted opinion pieces written by climate-change deniers. And that my conscience does not allow me to work for newspapers that give editorial space to climate-change or Holocaust deniers.

But *MittMedia* has no plans to change, so I say goodbye. And my last column is never published.

Scene 48. Unscientific Research

'New record!'

It's Saturday morning and Greta bounds into the kitchen waving a sheet of paper filled with figures and columns. 'Over one per cent are about the environment or the climate. Most are short pieces or old articles that are still up, of course, but still.'

It all started with someone we knew saying that reading the newspaper would soon be unbearable because there were so many awful things being reported all the time: 'It's just crisis upon crisis. War, Trump, violence, crime and climate change.'

Greta did not recognize her own experience in that description of reality, but our friend wasn't alone in her opinion: that there was so much awful news about the climate.

Our daughter – in contrast to our friend – thought that there was almost never anything written about the climate, the environment or anything else to do with sustainability, and for that reason she decided to take a good look for herself.

She started by regularly counting through what the four major daily newspapers in Sweden were publishing on their news sites – and what they weren't publishing.

How many articles were about climate and environment? And how many were about things that stood in fairly direct opposition to the subject – for example, package holidays, shopping or cars? The result was basically the same every time. Climate and environment varied between 0.3 and 1.4 per cent while the other topics were considerably higher.

*

When one of Sweden's largest newspapers started a climate campaign that would 'permeate the whole editorial office', Greta followed their reporting five weeks in a row and the result was less than impressive.

Shopping: 22 per cent; cars: 7 per cent; air travel: 11 per cent.

And the climate issue: 0.7 per cent.

Every single time she checked, whichever newspaper, the results were basically the same.

Greta is the sort of person who keeps track of things she thinks are important, so every morning we read all the newspaper headlines on a Swedish news-aggregator site.

'I'm going to make note of when the climate is the lead news story,' she says.

But no such note exists yet.

And we've been checking for two years.

Scene 49. Human Value

We're walking the dogs, strolling to the circular park behind Fleminggatan. Svante glances at the phone in his hand out of habit. The summer of 2017 is over and Moses has a new canine friend, a 'sister' we adopted six months ago.

Roxy is a coal-black Labrador who is at least as disobedient and loving as her big brother. If it hadn't been for the advocates at Dogs Without Homes Rescue she would probably have ended her days in a cage in southern Ireland. Instead she's walking here with Moses, happily nosing away on every single blade of grass. They never tire of it.

The past summer weather was – from a Swedish perspective – extremely average; we did not benefit one bit from the deadly heatwaves rolling through southern Europe. Even so, July was the second-warmest month ever recorded on earth. And here we were, experiencing another usual, mediocre Nordic summer. Typical.

On the other hand, for the past week our news has been filled by rainfall figures and flooding that are anything but usual. 'It's fake,' the climate-change deniers maintain on Twitter. But unfortunately there's nothing fake about the pictures of Houston highways turned into lakes ten metres deep.

It wasn't the 'usual' in Sierra Leone either. We look at a clip on the phone as the dogs continue nosing and tugging at the leash. In Sierra Leone they had three times the usual rainfall.

'Our house used to stand here,' says the man in the TV report playing on the little screen. 'We lived here,' he continues, pointing at a slope of red clay. The camera pans across what only a few weeks ago was a whole neighbourhood outside the capital, Freetown, but now there's no trace of it. No foundations, chimneys or wrecked cars. Only clay. Reddish-grey landslide clay.

The man says he misses putting his kids to bed in the evenings.

He misses singing goodnight songs to his son.

You see, he has lost all this.

His wife, his children, his home, and now he's going around in what little is left of his corner of the world, giving a British TV reporter a tour of the devastation. But there's nothing to show. Just a hill of red-grey clay and a handful of slow-moving aid workers in the background. Otherwise it's empty. Here had lived thousands of people. Families had their everyday routines here.

A life.

People who would wake up in the morning and have breakfast and send their kids off to school before going to work.

People like us.

The reporter is crying and doing his best to convey the man's fate, despite the fact that he presumably knows that this too is going to be drowned, although in a different sort of clay – a Western clay called the news cycle and the proximity principle.

He's trying to make a moving story, but the man from the Regent shantytown on the slopes of Sugar Loaf mountain in Sierra Leone does not seem the least bit eager to accommodate the crying reporter. He just stands there, expressionless.

Some allow themselves everything. Others allow themselves nothing.

Over a thousand people died on the slopes of Sugar Loaf in the wake of extreme weather. The man from Regent has lost everything and he doesn't even indulge himself to grief in front of the TV camera.

Scene 50.
The Proximity Principle

'Climate change is behind this,' the president of Colombia said when, in April 2017, he confirmed that hundreds of people had died in a landslide caused by unnaturally heavy rainstorms that had struck Colombia and neighbouring Peru.

But not many people were listening. And when the dizzying videos of rivers of metre-deep clay flowing through the village streets at 50km an hour – like lava after a volcanic eruption – were distributed, newsrooms in the Western world were only moderately interested. The videos got as little attention as all the other thousands of stories where people meet similar fates.

In journalism this is called the 'proximity principle'. And it means, for example, that a terrorist act that takes place in France is much bigger news than if a similar tragedy were to happen in Iraq, because Sweden is considered to have more in common with France than with Iraq.

It also means that when extreme weather phenomena take place, it takes a lot for them to become news if they don't happen in Europe, the USA or Canada – or Australia!

Because according to the proximity principle, Australia is considerably closer to Sweden than, for example, Lithuania, even though Lithuania is both a neighbouring country and a member of the European Union.

Different countries are assigned different value. Citizens in different countries are seen to be of different worths. Or at least different newsworthiness. But it can't be ruled out that the newsworthiness infects other values. Such as human values, for example. Or what one human life is worth. But what do I know? And weather is just weather, something which in a news context happens on its own. It has never been any different. Until now, when the world's scientists are drawing clear parallels between our emissions of greenhouse gases and the increase of extreme weather events that we're witnessing around the world.

Today you can read article after article in which leading experts explain that global warming works roughly like anabolic steroids when it comes to storms and extreme weather. Our emissions make extreme weather more extreme – there is a clear, accepted connection.

That connection must start having an impact on the news we choose to report, and how they are being reported.

Scene 51. Same Disease, Different Symptoms

When not a single line about the landslides in Sierra Leone was published in the Swedish media, we set about sharing the story on Twitter and Instagram. But before long we get drawn back into our everyday reality when the phone rings.

Greta is sad. She hasn't had any lessons the whole day, because no teachers showed up at school.

She still doesn't have teachers in several of her main subjects and we have to arrange yet another emergency meeting with the school administration. Greta is disappointed, because when she finally did get a good science teacher, she stopped teaching Greta's class because she wanted Mondays and Fridays off.

'It's supposed to be a school for children with special needs but it isn't,' Greta says with a sigh. 'It's a school for teachers with special wishes.'

So it's time to go home with the dogs and start calling around again, trying to get the everyday life to work for a little longer than one hour at a time. But the school principal is evidently in the Philippines, and no one there knows anything about the schedule having been changed four times in two weeks, or why.

'Drop it, otherwise you'll go under,' Svante says as I despairingly look out over the sunlit crossing on Fleminggatan. But I can't drop it, because if I do, someone else will have to take over and that person doesn't exist. I know exactly what he means but I can't let go, because it's impossible.

⋆

In the evening when everyone is asleep I sit on the sofa and cry out all the worry that I won't let the children see and that I have to keep inside.

The tears surge through my body and out through my fingers like a tidal wave of sorrow and all the fucked-up fucking shit.

All that frustration over never being able to let go of control. Then I remember all the emails that I haven't had time to send to inform all the teachers and school administrators about the situation in the school, and I write until my hands turn numb and the phone locks up and I lose feeling in my arm, and I hate everything, I hate myself, and I hate everyone.

I can't bear to explain any more.

Can't bear to ask for help.

Have to make fresh waffle batter for breakfast and arrange new melatonin and Oxazepam and call the doctor, who is on vacation, and the family has to quarrel about everything, and I have to be sad and the worry, worry, worry is like two tonnes of cement weighing on my chest, and I can't do any more.

It has to come out. I have to be able to breathe.

I lie awake and read about people who are much worse off than I am.

I read about all the burned-out people on a burned-out planet where weather, wind and everyday struggles intensify with every second of every day.

And I think that these are all symptoms of exactly the same disease: a planetary crisis that arose because we have turned our backs on each other. We have turned our backs on nature.

We have turned our backs on ourselves, I think over and over again until I fall asleep.

In a bed far from rain-drenched cities and the mudslides of Sugar Loaf mountain in Sierra Leone.

Scene 52. Killjoys

On 6 March 2016 I flew home from a concert in Vienna, and not long after that I decided to stay on the ground for good. It was necessary in an atmosphere where you couldn't take a position for or against anything without being faced with, 'I see, so what do you do yourself?'

Because our contempt for hypocrisy seems to be so monumentally great that we would rather sacrifice the only known form of intelligent life in the universe than affirm our imperfect good intentions.

It was a decision that was necessary in order to be heard. Because how can we begin to create the greatest demonstration of strength in world history if we aren't heard?

The issue of air travel summarizes the whole climate debate, the research is crystal clear, and yet we don't want to listen. But to stop flying is obviously not just about air travel itself.

It's about the fact that the world's species are dying out at a pace up to a thousand times greater than a 'normal' extinction rate.

It's about the fact that all our emissions have to be reduced to zero and then immediately onto negative figures with technology that hasn't yet been invented. And perhaps never will be. At least not at the scale required. It's about the fact that we lack sustainable ways of handling some of the extreme habits we've chosen to take for granted. Such as moving hundreds of tonnes of metal around the earth within the space of a few hours.

⋆

'This is my favourite,' Greta says, laughing out loud while reading through the comments section on a 'I've chosen to stop flying' article. 'If we're going to stop flying, the trains have to get better. Everyone says that! And what that means in practice is that the mere thought of a possible delay is so totally unreasonable that we would rather destroy the living conditions for all future generations than subject ourselves to that possible hold-up.'

Greta follows Roxy with her gaze. She pauses before continuing to speak.

'Everyone is so accustomed to the idea that everything should be adapted to their own needs. People are like spoiled children. And then they complain about us children being lazy and spoiled. I know that those of us with Asperger's can't perceive irony because it says so in the manual that all the old psychiatrists have written about people like me – but I don't think that irony can be described better than this.'

Scene 53.
'Like a Meteorite with Consciousness'

On Facebook there is a newly posted clip from a Danish news channel in which the host asks the studio guest whether it isn't a bit fanatical to stop flying.

'On the contrary, I think it's more fanatical to think that we can live with four degrees of warming,' the guest replies. 'What is truly fanatical is to think that we can continue living the way we do, with the norms that apply for the small elite we make up. So to stop flying is more like the opposite of being fanatical.'

Approximately 3 per cent of the world's population indulge in the luxury of getting on board an aeroplane every year. Even though flying is by far the worst thing you can do for the climate on an individual level.

The guest on the Danish TV news is not part of that 3 per cent. His name is Kevin Anderson and he is an adviser to the British government on climate issues. He is a professor at the University of Manchester, guest professor at Uppsala University in Sweden and previously headed the internationally renowned Tyndall Centre for Climate Change Research. And he stopped flying in 2004.

'Everything is about pies,' he likes to say. 'In order to limit global warming to two degrees Celsius we have a limited CO_2 pie, which contains all the CO_2 we can ever release. When we've used the whole pie there's no pie left. So that last little slice of pie we now have remaining must be divided up fairly between all the countries in the world.'

The idea of a common pie is as childishly simple as it is – in the true sense of the word – revolutionary. Because a budget sooner or later involves some form of rationing.

And within this idea we find the entry point to the end of the neoliberal world order that Margaret Thatcher and Ronald Reagan started almost forty years ago. It's not even a theory, it's basic preschool arithmetic.

The Dilemma, with a capital D, is that within this common pie our SUVs, our holidays and our meat-eating sit side by side with the construction of roads, hospitals and schools for billions of people who until now have done nothing to create the problems we are now facing.

And every time we choose to fly, eat meat or shop for new clothes it leaves a smaller CO_2 budget for less fortunate parts of the world. All according to Anderson's various lectures, which are accessible on the internet.

These are without doubt extremely difficult facts to relate to, but we can no longer indulge ourselves in looking the other way and pretending that the existential fork in the road doesn't exist.

The dizzying advance of modern society involves so many complications for the planet we live on – problems that by themselves would have been difficult enough challenges. The true worry is that we're doing everything all at once, at the highest speed possible. Humankind, Kevin Anderson says, is like a meteorite with a consciousness.

Scene 54. #IStayOnTheGround

Staying on the ground creates ripples in the water. And making ripples in the water is the best thing we can do – at least on all those days when there isn't a general election for the Swedish Parliament.

A friend asks me which flights are unnecessary. My flights, I answer. Just as unnecessary as my shopping and my meat-eating.

And no, no one maintains that it's going to be enough. No one believes that consumer power is the solution. But if my microscopic contribution can in some way hasten a radical climate policy, then I'm in.

But each person has their own life to live.

Each and every person has their hands full with their personal, impossible equations.

No one can demand that each one of us get involved in a crisis that no one is treating as a crisis. Such responsibility can never be placed on us as individuals.

Air travel brings everything to a head, but the growth society does not accept that the way forward might require taking a few steps back.

Forward is the only direction that counts.

Scene 55. At the Psychologist's

'What's the capital of France?'

I don't remember.

'What's the highest mountain in Sweden?'

I don't know.

'Who's the president of the United States?'

The year is 2016 and I'm undergoing a neuropsychiatric investigation with the psychologist. After hundreds of hours of reading, I'm fairly sure; after thousands of pages I've formed a pretty clear picture. Not only of my children but also of myself. But I want it in black and white.

Not because I think it's going to change anything, but I want to know.

If nothing else, maybe it can help the people around me. Although to be honest I don't care about that right now. I'm just so tired and sad, and I'm hoping that somewhere out there someone will have thought of something that will help me get out of bed in the morning. Some kind of pill, or whatever.

Something that will make my legs hold up. Something that will get me to see beyond this total hopeless darkness everywhere. So I fill out all the papers. I answer all the questions. For the thousandth time.

The psychologist talks away but I can barely hear what she's saying. Or actually, I hear her, but I can't formulate the answers. It's as if my thoughts get stuck. I want to ask for a glass of water but I can't remember the word for what you drink water out of. Glass.

It shouldn't be hard at all but it's like it's no longer there. The word has drowned in sound.

For me, everything is music and that's how it's always been; but I've always been able to turn it on or off as I wish, and I can't do that any more. The diagnosis takes over. I try to push my thoughts aside but the roaring of sound seeps in and soaks through, everywhere and all the time.

My gift and my curse.

My superpower, which has almost always been an asset to me, but as it is I can no longer control it, because nowadays all my energy is poured into trying to get everything and everyone to function.

'Who is the president of the United States?' the psychologist repeats, but the only thing I hear is that she is speaking in a monotonous minor G.

A window is cracked open and outside some birds are chirping in F9 with the third in the bass and the ninth in the four-line octave. It's out of tune. It's all a little sharp and it's bothering me so much that I don't hear what the psychologist is saying. It hurts. Physically hurts.

A motorcycle drives past on the street below in G, F, D, E, E flat, and it is much too low in relation to the birds' F9 chord. A creaking door, a notepad and a scraping chair form a cluster and my whole body screams in pain.

I really do want to ask for a glass of water. I swallow and blink in slow motion.

My fingers go numb and the psychologist takes a break and leaves the room. I say that I'll stay and check my phone but I just sit in the chair and shut my eyes.

Don't have the energy to stand up.

She comes back. Says it's probably ADHD and I'm showing

clear signs of depression and chronic fatigue. But the assessment is going to take a while. I drag myself to the pharmacy on the way home but they're out of most of the medications.

'The prescription isn't here,' the clerk says in a nasal voice between middle G, G sharp, A and B flat.

A zipper, a box being shut, a crying child and a truck on the street outside form a sprawling major 7th chord with the fifth in the bass. It disturbs me that the truck isn't rumbling in the tonic.

Beata's Theralen isn't available either and I can barely take this in. Without that we might just as well shut down our existence. Without that everything collapses.

'It's available in liquid form now. Have you tried the new flavour?' the clerk asks.

No, we haven't tried the new flavour and we haven't tried it in liquid form because it's more likely that Beata and Greta are going to learn to breathe underwater than manage to take medicine in liquid form.

'There should be a packet left at Kronan's pharmacy in Skärholmen.' But I don't have time to go all the way out to Skärholmen because Greta has just texted that the school staff threw out her rice because it didn't have a sticker with the date on it, which they have to do, but because of Greta's OCD she can't eat when she sees newspapers, papers and stickers – it's hard to mark her home-made school lunch with stickers, as we've explained a zillion times – and now Svante is on his way to Bergshamra to pick her up and I have to go home and make fresh jasmine rice.

But first I have to get the medicines.

I call an old friend, a now-retired doctor who has rescued me so many times in the past, but he doesn't have a computer and can't help me. I root around in my handbag for one of his old handwritten prescriptions and produce a pile of coins, the children's passports, receipts, hair ties and two pink dog-poop bags, but my fingers won't grip and the sound of everything falling back down into the handbag is like a gunshot in my ears.

The phone starts ringing at the same time as a text message jingles. Two emails. The sound cuts like a knife. I try to get the phone out to turn off the sound but my fingers still can't grip, it's like in my frequently recurring nightmare where I'm in the middle of a war zone and have to warn Svante and the children, but am incapable of typing a text message or bringing up their number.

My fingers cramp up.

Can't get the fucking phone out.

Try to open the screen lock with my chin.

Can't.

I walk out of the pharmacy over towards Willys to buy a snack for Beata and it's all about the air.

Breathing.

But there isn't enough air.

When stress levels increase, the oxygen intake is reduced, and even though I can hold a note for a full minute without needing to breathe, right now my lung capacity isn't enough to oxygenate my brain and muscles, and then I get even more stressed and then I take in even less oxygen and then it's even harder to think clearly and I don't want to be a part of this shit any more.

I'm standing on the pavement outside Västermalm's shopping centre and I am so dreadfully tired of all my hidden handicaps; all my invisible bloody problems. If only I could break a couple of bones. A fracture, a serious case of pneumonia or something else that forces you into a pleasant hospital for a few weeks so that you can get some sleep.

And breathe.

Rest.

Scene 56.
Dead Poets Society

There was a time when we caught the day with net and fishing rod – now we trawl the bottom of the ocean in our constant pursuit of self-realization, personal development and exciting new experiences. There are no limits. Everything is possible.

> Venice, the Maldives and the Seychelles are sinking into the sea, glaciers are melting, rainforests are being cleared and dry California is burning. Take the opportunity to visit these enchanting but climate-threatened places before they disappear for good.

These lines are too good to be true.

It's like something out of a Max Gustafson comic cartoon but reality, as we know, always exceeds fiction and in fact the quote comes from the front page of a 2018 issue of *Svenska Dagbladet*'s 'Perfect Guide'.

Climate tourism is a real phenomenon and it constitutes a significant source of income for people in many vulnerable places. But naturally for a limited time only. The coral reefs off Belize and Australia, for example, Mount Kilimanjaro draped in snow, and obviously the whole Arctic region.

Come and experience it before it disappears!

When Robin Williams's character in the 1989 film *Dead Poets Society* taught his students the meaning of the expression *Carpe diem*, almost an entire generation was watching.

He was a good teacher. And we were phenomenal students.

The Berlin Wall fell, borders were opened and the world was shrinking by the minute.

Airline tickets got cheaper, prosperity increased and suddenly the term 'weekend trip' was no longer only in the vocabulary of high-income people on Strandvägen in Stockholm.

Of course, not everyone could afford to endure a little jet-lag in exchange for the opportunity to go shopping in Manhattan over a weekend in October . . . but quite a few could.

Not everyone could afford to lie on a beach in Southeast Asia when the Swedish winter was at its worst. But quite a few could.

Quite a few more than we could ever have dreamed of when we left the cinemas that autumn, with Robin Williams's words firmly rooted in the back of our minds.

'*Carpe diem*,' said dear Robin, and we went out in the world and did just that.

But we did more than seize the day. We seized whole weeks, months and years. All in our hunt for cocktails at sundown, a new Danish-style kitchen, or a pair of shoes that couldn't be bought anywhere in the mountainous north of Scandinavia.

Reality always exceeds fiction.

Scene 57. Waffle Day

Over a year has passed since Greta's weight curve turned upwards again. Now she eats the same thing every day. Two pancakes with rice for lunch, which she heats herself in the microwave and eats during the break at school. She eats one thing at a time, never with any sauces or toppings. No jam or butter. What she eats has to be clean and she is extremely sensitive to tastes and smells. For dinner she eats noodles with soy sauce, two potatoes and an avocado.

Greta simply doesn't like to eat new things. On the other hand she loves to smell different foods. It's a passion. When she was feeling her worst she could spend hours going through the whole pantry, smelling every package, and on the rare occasions we eat out she sniffs right across the restaurant's salad or breakfast buffet.

One day a salesperson in the supermarket is offering tastes of waffles with whipped cream and jam. Greta goes up and smells her way through ten small waffles set out on the folding table.

'Now you have to eat them too,' the woman says when Greta is done almost dipping her nose into the whipped cream and jam.

Greta stiffens when the waffle woman confronts her.

'She has Asperger's,' I break in. 'And selective mutism. She only talks with her immediate family and she has an eating disorder, so she can't eat them. But she loves smelling things,' I explain and try to sound as friendly and apologetic as I can.

But the woman's face does not soften one bit. 'Then *you'll* have to eat them,' the woman says.

'Sorry. It really won't happen again.'

'Then *you'll* have to eat them,' the woman insists with such unexpected emphasis that I don't see any other way out than to finish every mini-waffle with jam and whipped cream on the spot while Greta waits at a strategic distance from the waffle woman, me and all the passers-by who watch what is happening with unforseen surprise.

We go out into Fleminggatan and I look at Greta.

She looks away.

'What?' she says. 'Smelling has to be allowed.'

Scene 58.
Co-autism

'It's important that parents don't adopt the diagnosis, because otherwise a sort of phantom autism can arise, and if you let the diagnosis take up too much space the problem is going to grow.'

Yeah, thanks for the update.

We've heard that warning since the beginning, long before we even suspected that the diagnosis was a diagnosis.

Many of our quarrels are about this. I want to challenge, explore and uncover. Preferably the day before yesterday. Svante wants to wait and give it time. It's like that in most families, the psychologists with whom we've been in contact say.

We understand all about co-autism, and it's true. But some days we choose not to understand. Some days we stick two fingers up at logic. Not because it makes things easier to handle.

It's just that some days we choose to play along with the diagnosis, because sometimes the diagnosis makes sense and the norm does not.

Scene 59. Tick Tock

Nothing is black and white. The world is complex.

There are always different truths and in an open society all sides must get the chance to be heard equally. That impartiality, which is the basis for the democracies in our part of the world, is in many ways ingenious. Except when it comes to the few cases that are in fact black and white.

Like life or death.

Or the questions where the 'grey zone' is intertwined with such great risks that any debate ought to be ruled out.

The climate and sustainability crisis is far from uncomplicated or simple.

But in many ways it is black and white.

Because either we make the Paris Agreement's well below 2°C target and avoid the risks of setting off catastrophic chain reactions far beyond our own control – or we don't.

That difference is as black and white as it gets.

There is even a clock counting down the time we have left before it's too late to achieve the 2°C target. The clock is based on official UN figures and at the time of writing it stands at 16 years, 355 days, 13 hours, 22 minutes and 16 seconds. There is a similar clock counting down the remaining time at our current emissions rate for the updated IPCC 1.5°C target and that one stands at 7 years, 11 months, 17 days, 14 hours, 22 minutes and 16 seconds. And that gives us a 66 per cent chance of staying below 1.5°C.

At the time of writing, a number of leading researchers estimate that we have an approximately 5 per cent chance of making the 2°C target.

Scene 60.
'What's this about again?'

Beata doesn't want to go to P.E. because at P.E. you have to throw hard balls at each other and the balls hurt. She doesn't want to go to P.E. because there you have to play all different kinds of sports that are based on defeating each other; sports that all the boys seem to love and that are full of screaming and jostling. She doesn't understand why they can't dance instead, if exercise and co-ordination skills are what's important.

Beata dances all the time at home, but she never gets to do it during P.E. class.

Beata doesn't want to go to woodworking class either, because she's scared to death of the machines, and she doesn't want to play cards during breaks because no one understands her rules where the queen beats the king.

'And why should boys always be worth more than girls? Why does everyone always have to laugh at the boys' jokes and why is everything about being seen and heard when it's always the boys who are being seen and heard the most?' Beata asks and turns towards me. 'Mum, what's this about again?'

'The patriarchal structures of society,' I answer.

Scene 61. Moscow Pride

In the hours before the Eurovision final in Russia in 2009, a Pride parade was held on the streets of the host city, Moscow. It was a radiant spring day and we performers were following everything via social media from inside the arena. The dress rehearsal was about to start when the news spread that the Russian police had broken up the parade and that around eighty participants had been arrested and taken away.

Everyone knew. No one backstage was talking about anything else.

'The parade damages society's morals,' the responsible city officials announced.

It was our audience which had been dragged away by force outside the arena, and to me it felt obvious to express my support for them and my loathing for the Russian authorities.

'Shame on you, Russia,' I said, thinking that we couldn't very well make an entertainment programme while part of the audience was in jail for standing up for fundamental human rights.

But of course we could.

In the end it was only Spain's contestant, Soraya, and I who expressed solidarity with our imprisoned supporters; everyone else was strategically unaware. Strategically uninterested.

'No politics in Eurovision,' they all said. As if the rights to love the one you love were political.

The jury placed Spain second last and me third from last.

And that sunny Saturday in Moscow was one really shitty day at work.

When it was all over, press conferences and exclusive interviews with the Swedish media on board the performers' bus awaited. The following day I would once again sing *La Cenerentola* at the Stockholm Opera, and all I wanted was to get the hell away from Moscow. Home to the kids.

'Make sure not to look sad now,' everyone said. 'And don't cry or look disappointed until the bus is well out of view of the photographers.'

'And don't say anything about being disappointed,' someone added.

I understood and did exactly as they said. The fact that we found ourselves in some kind of a dictatorship that imprisoned homosexuals no longer mattered, of course – now it was all about not looking like a loser.

Now it was all about not crying. Not looking weak.

Scene 62. Digital Success

'No, don't respond. Then you're going to sit there arguing all evening with some Russian robot troll that's programmed to tire out people like you.'

Greta is logged in on one of her animal-rights accounts on Instagram and giving her favourite arguments to her favourite antagonists. The climate-change deniers. The techno-optimists. And especially the frequent-flying vegans who traverse the globe in order to 'save the world' with new, exotic recipes. She looks content.

'There!' she says, pleased, opening her eyes wide. 'That showed him.'

'But you shouldn't engage,' Svante says. 'It's a waste of time. What did you write?'

'It was a pilot who's a vegan because of animal rights . . . as if the animals don't need a functioning atmosphere too?' our daughter replies. 'And he said that the climate crisis was because there are too many people.'

'Okay. Did you respond as usual?'

'Hmmm . . .' Greta nods, smiling with her whole face.

She has a few stock responses saved in Swedish and English, and one of them is about the 'population problem', a constantly recurring argument:

> Our emissions are the main problem. Not the people. Most humans live well inside the planetary boundaries, unlike us in the global north. The richer you are, the greater the emissions. So if you want to limit the population to save resources,

> you ought to start a campaign to rid the world of billionaires. Why don't you call it 'Get rid of all Bill Gates' and prohibit all CEOs and movie stars from having children!' But it would probably be a little difficult to get the UN to adopt such a resolution, so I recommend that you reduce your own emissions instead. Or support education for girls in the global south because that is the most effective way to limit increases in population.

'What was his response?' I ask.

'Nothing,' says Greta. 'Or wait . . . he blocked me,' she says, and laughs so loud that Roxy jumps up on the couch and starts barking.

Scene 63.
Hubris

In historical terms, we are facing unprecedented changes in every single aspect of our global community.

But leaving that 'endless growth' society, which has given us so much and raised so many of the earth's population out of poverty and misery, is easier said than done. We are still intoxicated by the incredibly successful fairy tale of progress that transported much of the western world from starvation and misery to lunar landings, 24-hour entertainment and retirement homes in the southern sun.

In three generations we have gone from exposure and vulnerability to immortality, and we sometimes act as arrogantly and short-sightedly as if we were secretly stranded on an unchartered desert island, far from any shipping routes, with provisions that could last a year, but which we stuff ourselves with in the first week.

'Just, rely on technology – someone is going to find a solution,' everyone shrieks in chorus as they toss rubbish in the freshwater spring and set fire to the lifeboat so as not to be cold at night. 'You only live once. Enjoy!'

It was reduced inequalities, collective solutions and a dawning humanistic worldview that raised us out of poverty. We peeked through the door to fairness and equality but now it's slowly closing again. The gaps increase, resources dwindle and we are stranded on a desert island in the cosmos.

Scene 64. Retake

'Okay, so how about this,' says Greta. The spring sun is shining and we're sitting together at the summer house on Ingarö and realizing that what we truly want to say with this book is almost impossible to formulate.

'Feminism is standing outside one door stamping its feet, eager to get in. The door is locked but you have to get inside in order to move ahead. A little further away are the other movements – humanism, anti-racism, the animal-rights movement, those who fight for refugees, or against mental illness, or economic differences and so on. Everyone is standing by their own door and wants to get in and move forward. The climate movement has a key that fits all the doors, but no one wants to accept help from it. Either they're too proud, or else they don't see that the solution is right in front of them. Or they don't want to lose all the privileges that the climate movement opposes.'

'Okay,' Svante says. 'Say that again, exactly the same, and I'll write down every word.'

Scene 65. Greenwashing

According to a recently published study by the UK organization Influence Map, forty-four of the world's fifty most influential lobbying organizations are actively working against an effective climate policy.

It's actually extremely simple.

We are entirely dependent on companies' efforts and willingness to find sustainable solutions. But we can't turn all the responsibility over to them. It's neither just nor reasonable.

The primary purpose of a corporation is, after all, to produce economic profit. Not to save the world.

And all the claims that there is no contradiction between these two conflicting goals will always ring a little bit false, and it is right here that we stumble on the phenomenon of 'greenwashing', in the gap between beautiful words and real action. In their business strategies. Or tactics disguised as new technology.

Nothing exemplifies that phenomenon more clearly than Naomi Klein's description of the entrepreneurial guru, airline owner and billionaire Richard Branson's profitable adventure as a philanthropist and climate hero in her 2014 book, *This Changes Everything.*

A little over a decade ago, Al Gore gave Branson a private climate-crisis presentation. Branson was so moved by what he heard that shortly afterwards he called a press conference where he announced that during the next ten years his company would invest $3 billion in finding a way to produce sustainable jet fuel.

Because his business operation had earned such vast sums of money from services that necessitated great quantities of CO_2 emissions, Branson thought it only right that he invest part of his earned and future profits in finding a solution to the problem of aviation's climate effects.

And not only that.

Branson also ran a competition where anyone who discovered a technical solution for soaking up such and such amount of CO_2 from the atmosphere would win $25 million.

This was amazing news. Not least for people who relied on flying to be able to do their jobs. People like me. It felt like everything would work out just fine, because this was one single company and if it was as easy as this then everyone else would be investing too. Not to mention all the governments of the world.

It felt soothing.

And good.

There was a solution!

The problem was, Branson never found a new fuel that was sufficiently sustainable to fulfil the requirements. He didn't even get close.

The closest solution was biofuel, but the problem with biofuel is that there isn't enough forest and agricultural land to cultivate the quantities required. And of course, the rainforest is already being cleared at horrendous rates and the number of non-tropical countries with extensive forests such as Sweden, Finland, Canada and Russia are limited. There is simply not enough forest available.

Besides, biofuel is expensive. And then there was that moral conflict over the fields being needed for other things. Food, for example. Not least for the 89–90 per cent of the world's population that had never set foot on an aeroplane at that time.

In the end, instead of $3 billion, only $230 million was invested.

During the years that followed, Branson started another three airlines and a Formula 1 team instead. No winner of Branson's Earth Challenge prize of $25 million has yet been announced.

'Green' aviation is like Donald Trump's 'clean' coal, or so-called carbon capture and storage (CCS). It sounds good, but it's not going to work in time. Except for the companies involved, of course.

The same companies that say that everything is going to work out – all we have to do is keep buying their 'green' products.

Scene 66. Ski Outing While Waiting for Time Machine and Teleportation

It is a brilliant winter day and we are on our way out across the ice in the bay. We've bought a pair of second-hand cross-country skis for Greta. Her little sister has stayed at home in the city. Beata likes to be by herself and transforms the whole apartment into a concert hall where she rehearses, acts, sings and dances.

There she feels her best.

She's preparing material for her YouTube channel, which she will launch when she's ready.

'But absolutely not for two years yet. It has to be good first.'

So we split up as often as we can.

Svante goes first and Greta and I each hold on to a leash pulled by Moses and Roxy as fast as they can go. We're moving really quickly. We can barely stand up, and we shriek and laugh at our speed in the wind.

We ski along Björnön and almost fly past the paths and beaches and cliffs that are packed in glistening, crusted snow and overturned winter ice.

Once at the inlet we sit on someone's dock in the sunshine eating oranges. Svante peels, I eat and Greta smells. It's a good day for the Ernman–Thunberg family.

A short distance away we see three families out with their All Terrain Vehicles teaching their small children to drive their own, gasoline-powered children's models. Each family has three ATVs.

'Look at that,' Greta says, 'a family that shares an interest in engines and motorsports. How nice.'

I snort with laughter. The orange slice flies out of my mouth.

The children are six or seven years old at most and it wouldn't surprise me if their parents are the very ones that spend the summers teaching their children to ride back and forth on their own little jet skis and compete with Dad, up and down the bay.

Back and forth.

'So nice to see that they're out and compensating for those who took the bus here or spent half a fortune getting an electric car,' Greta says, smiling and pushing the ski a little with her foot.

The best thing about being a slightly upper middle-class proponent of technology is that once you've acquired an electric car, solar panels and a Powerwall, you realize rather quickly that technology isn't going to solve everything. Because changing your habits does far more than virtually all technical solutions when it comes to reducing your own emissions. Both are needed of course, but while waiting for CO_2 vacuum cleaners and time machines there are two things we need more than anything else: radical politics and new legislation.

Because for every new electric car there will always be a new jet ski. For each person who starts taking the bus there is a new petrol SUV. For every vegan there is another imported Brazilian fillet of beef. And for every person who refrains from flying there is a new weekend trip to Madrid.

Consumer power can influence public policy, but it's not a solution.

Four years ago we had the opportunity to install an electric car-charger in our garage, and we replaced our fossil-fuel car with an electric one. Two out of sixty car owners in our building took the opportunity to switch to a pure electric-powered vehicle immediately, while one person invested in a rechargeable hybrid. Since then, quite a few new cars have appeared in the garage. Many in the same price range as ours.

But not one new electric car.

Not one new plug-in hybrid.

Same thing with the solar panels on the roof.

They've been up there for four years.

And for four years we have enthused about the new technology. But none of our neighbours have followed our example, and naturally it's no different elsewhere in the world.

The solutions exist and they work just fine. With renewable energy sources like solar and wind power we already have the ability to start a rapid phasing-out of the fossil-fuel society. Technology moves ahead but the investment is too slow. Much too slow.

Everyone seems to believe that technology will rescue us. But the energy companies put the brakes on development and we private individuals who have the chance to push the development forward don't seem the least willing to use that very same technology we so faithfully claim to believe in. Or rather: we don't seem to believe that we need to be rescued.

The Ernman–Thunberg family leaves the ice and plods home through the headwind.

Scene 67. Greta's Monologue

Greta is sitting on the floor in the kitchen with Moses and Roxy. She's brushing their coats with an old comb, calmly and methodically.

'I remember the first time I heard about the climate and the greenhouse effect,' she says. 'I remember thinking it couldn't be true. Because if it was, we wouldn't be talking about anything else. And there was basically no one saying anything about it.'

'You're the ones who are going to save the world,' I tell my daughter.

She snorts just like my father does, a man who has gone through life like a spot-on caricature of a person with Asperger's syndrome. Undiagnosed. They are so laughably alike.

'All the teachers say exactly the same thing,' Greta replies. ' "Your generation is going to save the world. You're the ones who are going to clean up after us and fix everything," they all say, before flying off on vacation on every break. "You're the ones who are going to save the world!" Yeah, so you keep saying. But rather than saying that, we children sure wouldn't mind if you started helping out a little bit instead.'

She gets up and follows Moses, who has settled down on the rug a few metres away. She continues.

'And no, Mum, people like me are not going to save the world. Because no one listens to people like me. Maybe we can learn things, study and educate ourselves, but that doesn't count any more, does it? Just look at the researchers. And the scientists. No one listens to them. And even if they are being listened to, it doesn't matter, because the companies who destroy our nature and our climate hire lots of their own "experts" that they send to the USA for the world's most super-expensive media training so they can be on all the news channels and say that, no really, cutting down all the trees and killing all the animals is a good thing. And when the scientists disagree they aren't heard because the companies have already plastered half the country with ads telling their side of the story; because the truth is just another one of those things that can be bought for money.'

Moses lifts his head when the elevator starts in the stairwell. Greta follows his gaze.

'You have created a society where the only things that are valued are social competence, appearance and money. So if you want us to save the world you'd better hurry up to change the society that we all live in. Because as it is now, everyone who thinks a little differently and who comes up with original, out-of-the-box ideas that don't occur to other people falls apart sooner or later. Either they get bullied in school because they're different or else they sit at home in order to survive. Unless of course they manage to get into a special-needs school like I did, but then again we don't have any teachers because hiring teachers decreases the school's profits.'

She turns round and looks me right in the eye. She almost never does that.

'You always keep reminding me that I got a big publishing company to promise that they would rewrite the high-school geography textbook, after I pointed out that what was in it was wrong. They even wrote an article about it in a journal on sustainability.

'I haven't had science in almost a year. Because we don't have a science teacher. So if you adults want a future for this world then you have to change it real fast. Because as it is now, absolutely nothing is working.'

Greta takes a deep breath and buries her nose in Moses's thick white coat. Sniffs.

Scene 68. It Wasn't Better Before

Less than a hundred years ago it was still an accepted truth that certain countries had the legal and moral right to control and own other countries – that right was as obvious as the fact that not all people were of equal value, depending on their origin, skin colour, religion, sexual orientation, economic background or gender.

Many injustices have disappeared, many remain. Some have changed form and new ones have emerged. But most things have got much better.

The problem is that most of these improvements have been at the expense of other things, things that are not simple to repair or replace. Like mental health. Biological diversity. A well-balanced biosphere. Abundance of species. A clean environment. Breathable air. And climate stability.

Scene 69.
Goethe's Faust

Another thing that hasn't improved is the concentration of CO_2 in the atmosphere.

The connection between our historic journey to unprecedented prosperity and the amount of greenhouse gases in our atmosphere is undeniable.

Dust thou art, we said, *and unto dust shalt thou return.*

All that once lived shall rest in the earth, we said.

Then someone found a whole lot of oil and that was the end of all that resting in the earth. Instead we created a society based on digging up fossil remains to be burned up in the planet's highly sensitive atmosphere.

And how we fired away!

According to a 2003 study by Jeff Dukes at the University of Utah, 23.5 tonnes of biomass are required to produce 1 litre of petrol. That's 23.5 tonnes of old trees and dinosaurs, plus tens of millions of years so that a single Volvo can be driven 10km.

Much can be said about the contract our modern society has signed with the planet we live on.

But sustainable it is not.

Scene 70. Healing

The planet is suffering from serious disease and we need to initiate extensive medical protocols immediately. We need urgent care.

But instead – in the best case – we have chosen faith healing as our treatment method.

There is no insight into the illness. Not a trace.

It's like refusing emergency surgery for the option of someone inventing a magic, future cure.

Scene 71. London

Two dozen British eight-year-olds have spontaneously formed a children's choir around the karaoke salesman on the fifth floor of Hamleys toy shop in Regent Street.

They're singing 'Shape of You', by Ed Sheeran.

Beata and Svante are in London for her Christmas present from last year – to see her idols, Little Mix, at the O2 arena. In the year that has passed since we bought this present, Svante has changed some of his habits.

He has stopped flying.

Like me.

At first we thought it might be good if one of us kept the option of being able to jump on to a plane, in case of any unforseen emergency.

But then Svante read *Storms of My Grandchildren* by James Hansen, who was the head of NASA's Goddard Institute for Space Studies from 1981 to 2013. After that he read another twenty-odd more books on the same topic and then it was goodbye to shopping, air travel and meat for him.

So a quick round trip to London for a few hundred kronor was thus transformed into a significantly longer and considerably more expensive little adventure. Beata's Christmas present had been struck by a kind of moral hyperinflation. But a promise is a promise.

Our younger daughter has nothing against becoming a climate pioneer and will happily spend five days travelling in an electric car through Europe with Little Mix turned up to the max on the stereo speakers.

At Hamleys, Beata buys a fox as a Christmas present for Greta and then they stroll under the angel Christmas lights towards the HMV shop opposite Selfridges. Svante takes a photo of Beata at Oxford Circus and sends it home.

An hour later I pick up the phone. The screen shows two messages: the picture of Beata at Oxford Circus and a news-flash: 'Terrorist Attack on Oxford Street'.

I call and they answer immediately. They are already back at the hotel, far from Oxford Circus, and I can relax. An hour of extra news broadcasts and direct reporting on all channels follows. For a few minutes the world stops and everyone listens. Everyone watches. Swedish tourists are interviewed via mobile phone and everything is chaos, no one knows anything and everyone holds their breath.

But it's a false alarm. The police and military have been called out and everyone has done everything by the book, but nothing has happened except possibly a fistfight and someone who chased someone who chased someone, and now the Christmas shopping can resume. The world can calmly keep on consuming itself into a coma.

The next morning Beata stays in the hotel room. Singing and dancing are infinitely more exciting than exploring the world outside – no city in the world can compete with that. She is more than content to take care of herself, so Svante spends the day walking around the luxury yachts in St Katharine Docks.

There he sees privately owned motorboats with names like *Sand Dollar*, each and every one so big that it could be used for an international ferry service. He wanders along the shipyard and the piers by the Thames that once were the beginning of the modern world. Here were the trading companies, and here the ships and goods came in. This was the very heart of it, Svante thinks.

Trade goods that built the empire that laid the foundation for the Industrial Revolution. Trade goods and industrialization that kick-started the unnatural pace of the greenhouse effect. The same greenhouse effect that was discovered by the Thunberg family's own Nobel prizewinner, Svante Arrhenius. The man he is named after, because it was a big deal to be distantly related to a Nobel Prize winner, even though no one in the family had a clue about what he actually got the prize for.

My Svante walks around and reads on his phone that Svante Arrhenius's calculations about temperature increases – in his 1896 book *Über den Einfluss des atmosphärischen Kohlensäuregehalts auf die Temperatur der Erdoberfläche* – match pretty well what we know today, more than a century later. In fact, very well. What does not match, however, is the timescale.

According to Arrhenius's calculations, it should have taken up to 2,000 years for the CO_2 concentration in the atmosphere to be at the level it is today. Of course, he could not easily predict that future generations would dope themselves with all those fossil fuels that perhaps, to some extent, should have been kept in the ground.

Hour after hour Svante walks around, surrounded by hordes of tourists from all corners of the earth. Young, elderly, poor, rich: people from all sorts of backgrounds saunter around the Tower of London in the late autumn sun and post countless global selfies on all conceivable media platforms. A whiff of burnt almonds and diesel fuel from the passing sightseeing boats mixes with the too-warm November air.

Some of the tourists are so old that they don't have the energy to keep walking. Others hop along on crutches. American families with small children console their newborn infants.

A woman from Australia leads her obviously senile husband around and points out the skating rink by the curtain wall, where some Brazilian tourists wearing Santa hats are staggering around the ice in the perfect 18°C skating conditions.

La dolce vita. The sweet life. You only live once. *Enjoy!*

Svante sits in the sun by Tower Bridge under trees that still have all their leaves even though it will soon be Advent. He dreams about a climate movement that doesn't exist; that cannot exist yet, because it must be created by everyone as we go along.

He has a Starbucks latte with an extra shot of espresso, eats dry saffron buns that Greta sent with them, and pays the electricity bill to the Swedish utility Vattenfall on his phone.

The same government-owned Vattenfall that the year before sold coal mines to Czech venture capitalists who believe in a 'renaissance for coal power'. The same government-owned Vattenfall that still imports millions of tonnes of coal from heavily criminalized mining companies in Colombia to northern Europe, where it is burned up in dirty coal-fired power stations.

The same government-owned Vattenfall that is 112th on the list of the 250 companies that emit the greatest quantity of CO_2 – companies that, combined, account for 30 per cent of the world's emissions of greenhouse gases.

The same government-owned Vattenfall that sued the German government for billions because after the Fukushima catastrophe they chose to start phasing out nuclear power.

The same government-owned Vattenfall whose chief operating officer – certainly as competent and congenial as anybody – had recently become chairman of Sweden's climate policy council.

Svante takes off his sweater in the sun. It's T-shirt weather and a bird is chirping on a plastic lawn.

Scene 72. The Long Way Home

Beata starts crying as soon as she sees the name Little Mix on a billboard in the Underground.

'I'm not made of stone,' she sobs.

And when Little Mix rise on to the stage through four holes in the floor at London's O2 arena both she and Svante are screaming. No one screams louder and no one cries harder than Beata. No 'Mixer' in the whole world sings along to every key, to every harmony and to every line the way she does.

After the concert they get into the electric car and drive towards the tunnel to Calais and Kungsholmen. Beata is inexhaustible. She sits in the back seat and eats cookies and listens to all the albums at full blast. As long as it's Little Mix she has no sound sensitivity – then it *has* to be loud.

Come evening she watches *Friends* on the computer. She sleeps by herself in a separate hotel room and she is quite happy and perfectly calm amidst the chaos, because it's her choice to be there. She enjoys it, and as long as the car is moving she feels great.

Outside Eindhoven the phone rings; it's a publisher wondering if Svante and I want to be involved in a book about the climate. It's going to be a generous and optimistic book sold at a deep discount in order to reach as many readers as possible. The publisher describes our imagined role, and how important it is to publish something together about the environment that can reach a great many readers.

'It should be aimed at a broad target group and be full of hope.'

'Mmmm,' Svante replies and Greta's words echo in his ears: *A single flight can erase twenty years of recycling* . . . 'But we're probably not that interested in a book about hope at this point. At least not what is generally understood as hopeful right now.'

'What do you mean by that?'

'We don't think that hope is what's needed most now. That would be to continue looking away from the most important parts of this situation. If we're going to do a book about the climate, we must first and foremost communicate that we find ourselves in an acute crisis, and what that crisis involves. Hope is extremely important, but it will come later. When your house is on fire you don't start by sitting down at the kitchen table and telling the family how nice it will be once you've finished renovating and building the add-ons. When your house is on fire you call 999, you waken everyone you can and you crawl towards the front door.'

'Yes, however, I do think we need hope,' the publisher says. 'Did you know, for example, that if we simply adjust the air pressure in all tyres we would save over 100,000 tonnes of CO_2?'

'Sure,' Svante answers, 'but that's not something we want to focus on. If people think that such simple things can really make a difference, then we're making it seem like people can carry on as per usual. Pumping up tyres is really good, but that's a drop in the ocean, and if we devote the little focus that the climate issue gets to such things, then we're done for.'

'But if people think we're done for, they're just going to give up. They will think its game over.'

'I don't believe that at all,' Svante says. 'They won't, not once they are well informed about what the word "over" actually entails. Because they don't know. People unfortunately have no idea what a runaway greenhouse effect is. Or how close we've come to setting in motion things that can't be undone.'

'But there are psychologists who say that will make us shut down. Purely as a defence mechanism.'

'Yes, but there are psychologists who say the opposite too, and so what's the alternative?' Svante continues. 'Lying? Spreading false hope? Is that the way we'll get people to change their attitudes? Malena and I aren't doing this because we have a low opinion of people. On the contrary, we're doing it because we love people. Because we believe in people.'

'Okay, but what do you say to your neighbours when you talk to them?' the publisher asks.

'I don't talk to my neighbours. I can't even bear to talk with my friends or my own parents any more.'

The publisher says that he'll call back later.

But needless to say he doesn't.

Late in the evening, queuing for coffee at a McDonald's at the Hamburg Süd truck stop, Svante tells a man in broken German that he is on his way from London to Stockholm in an electric car because he has stopped flying *für das Klima*, and although the man understands what he's saying, he doesn't understand what he's saying, and there in the car park in the wind and the rain, Svante cries openly for the second time in fifteen years.

Because there, surrounded by 50 billion lorries, motorways and BMWs, he realizes that it doesn't matter how many electric cars we acquire.

It doesn't matter how many solar panels we put up on the roof. It doesn't matter how much we encourage and inspire each other.

And it doesn't matter if we stay on the ground and renounce

the privilege of flying, because what's needed is a revolution. The greatest in human history. And it has to come now.

But it's nowhere in sight.

For five minutes he stands there, until he realizes that no one can live with the thought of giving up. And that nothing will be solved by crying at German petrol stations.

All there is to do is drive on. Towards Jutland.

Towards Malmö.

Towards the dawn.

III

The Ancient Drama

'I walk past the headlines about murder and celebrity parties, about bans on begging and manoeuvring by newly elected party leaders. And I think: we're all sitting in a car heading straight for a rock face while we're arguing about the music we're going to play on the car stereo.'

– Stefan Sundström, *Dala-Demokraten*, 27 October 2017

Scene 73. Chaos

I love chaos.

I love the impossible. All those things that no one else can manage.

Doing cartwheels, handstands, or hanging upside down in a harness over the stage. Doing push-ups while singing an aria that's a struggle to sing standing still.

And I'm at my very best when there is a power cut before a live TV broadcast so that no one has time to rehearse and you simply have to solve everything on the spot in the studio. When I'm asked to cover for someone with three hours' notice to sing a role that I haven't sung in eight years and the performance is to be broadcast live in every cinema in Sweden.

Or when someone is ill and I have to jump on a plane to sing at a sold-out concert for 2,000 people in London and I'm handed the score once I've landed at the airport and learn everything by heart in the taxi on the way to the Barbican.

I love chaos.

As long as the chaos is mine and I get to do what I'm good at. That's when I'm at my best.

I have ADHD and of course I've had it my whole life.

I was forty-five years old when I got my diagnosis and the reason I hadn't looked into it sooner was that I never had any major problems that gave me reason to suspect that it was necessary.

I embody the 'superpower' that everyone talks about. The one which is often mentioned but which unfortunately only a few possess, because only a few happen to be in the right circumstances.

I can hear all the instruments in a symphony orchestra at the same time. And I can see their parts in front of me when I hear them play.

I was lucky. There were people who very early on made sure that I ended up in the right context: surroundings that happened to perfectly suit me, my talent and my peculiarities. Surroundings where I could devote all my time to what I loved.

I was shy and I stammered so much that I had to go to a speech therapist for several years in elementary and middle school.

'Are you going to the sp-sp-sp-sp-speech therapist again now?' the boys would jeer when it was time for me to leave the classroom. I couldn't say sentences that started with vowels and it took so much energy to talk to people that I preferred to stay silent.

But when I sang, everything became so simple and clear.

It was my salvation: in song I found my place on earth; there I was safe. There were no limits. I could be there for fourteen or fifteen hours a day and just sing, listen and write down all the parts, all the notes, all the sounds.

There was nothing I couldn't manage.

And today I still have that place inside me, a kind of feeling located in the muscle memory. A happy feeling that is mine alone.

When I sing I'm always happy.

Scene 74. The New Currency

Our collective ignorance about the climate and sustainability crisis has become one of the world's greatest economic assets. You see, that ignorance is a prerequisite for continued business as usual. Ignorance is our new currency.

Because the moment we realize the extent of the sustainability crisis, we will have to change our habits and take a few steps back. Such insights do not benefit an economy based on us continually filling our cars and aeroplanes with the remains of ancient fossils, while manufacturing and buying as many objects and items as we can as fast as possible.

The connection between growing economic prosperity, increased emissions and lost biological diversity is as clear as day. But that connection doesn't reach us. It gets drowned along the way.

Because our ecological ignorance has suddenly gained an astronomical economic value, and the result of this is visible everywhere: in the media, in news reporting, in advertising, in our values and habits. We don't have time to separate what has true meaning from all that is constantly distracting us. So everything just continues.

But the problems we're facing aren't going to disappear by themselves, and the costs of putting everything right again are simply going to increase by each day we delay.

Taking those steps back is inevitable. The question is, will we take them now, while we can still do it in an orderly way – or wait until later . . . ?

Scene 75.
Social Animals

'Climate change is the greatest threat facing humankind,' UN Secretary-General António Guterres said on 29 March 2018.

We have started a 'destabilization process' and we are moving towards what is called a 'tipping point', a point of no return which could be anywhere along the road we're currently travelling on.

An invisible boundary.

The fundamental debate should have been resolved long ago. Because the research has established, with all possible clarity, that global warming will lead to catastrophic changes for almost all living species. That deforestation, industrial agriculture, ocean acidification and over-fishing contribute to eradicating biological diversity.

But we live in a time when an increase in car sales can still give us faith in the future. A time when a flight delay can generate considerably more headlines than the deaths of thousands of people as a result of climate changes that have been created in part by our air travel. We live in a time when a 'climate-smart tip' is to replace teabags with loose tea.

I read that when you fly you should pull down the blind during take-off and landing to conserve the fuel that runs the plane's air conditioning. In the hotel room I can 'save the world' by hanging towels up on the hook instead of sending them to the laundry every day.

'We can't digest all the negative reports and all the dark news. Our defence mechanisms tune out the reporting. We need a new, positive story,' say those with a voice. And I kind of wonder, what *old* story is it that everyone seems to know about and that now needs to be replaced?

Because hardly anyone I know is able to keep any track whatsoever of the ongoing sustainability crisis taking place around us.

Hardly anyone we meet has the slightest idea about 'forcings' and 'feedbacks' and how a displacement of ocean currents under the Antarctic ice shelf can hasten the melting process. No one we're acquainted with knows that the same deforestation that we are upset about in the Amazon is taking place right around our own corner in the northern forest belt. No one we talk to has heard about the new Pangaea, or about the two companies outside Zurich and Vancouver that are furthest along in their work on a new technology that will vacuum CO_2 out of the atmosphere. And there is definitely no one we know who has read their business plan, taken out a calculator and concluded that this is never going to work in time.

In fact, we hardly know anyone who knows a single damned thing about the climate crisis. But that's not something to be ashamed about; we've met sustainability managers and party leaders who know just as little.

The fact is, we all lack the fundamental knowledge we need to be able to understand the crucial changes that follow in the tracks of our lifestyle.

I myself knew nothing about the climate issue three or four years ago. I was a little worried of course. I thought that our habits must be taking a gigantic toll on the earth's resources.

But every time I read about something that was damaging, there was always someone who maintained the opposite, and it felt incredibly reassuring that the reporting always offered such professional refutations to every conceivable ecological concern.

Everywhere I was met by the same happy message: the solutions exist, go ahead and carry on as usual! I read about aviation. Apparently it's extra harmful because the emissions happen at high altitude. But neither the Swedish Civil Aviation Administration nor government-owned Swedavia, which runs Sweden's airports, said anything about such damage.

Visiting their websites, you were met by pictures of air-traffic control towers among blossoming tulips, and beautiful words about green transition.

And so it was with everything else too. If there was anything wrong, technology would fix it. As if *global warming* itself were the fundamental problem, as if the climate crisis was not in reality a symptom of our over-consumption and our unsustainable lifestyles.

But I had no problems with that.

As long as the media and politicians didn't give off the impression that anything was seriously wrong, I assumed that everything was under control.

Then came Greta's crisis, followed by Beata's, and it was as if we stumbled into a room we didn't know existed.

The idea that 'We need a new story' feels more and more peculiar. It assumes that everyone has seen *Before the Flood* with Leonardo DiCaprio and then kept on researching on their own, reading science reports and climate blogs.

It assumes that everyone regularly listens to TED talks and reads the *Guardian* meticulously, and that we all understand the full significance of the sustainability crisis.

'We can't bear to take in all the negative reports. We have to think positively, because otherwise we shut down,' say those who are supposed to be the experts. But that is not true. Because we can't repress something we don't know anything about, and we can't ignore what's not being reported.

A parent whose child has strayed towards the edge of the cliff doesn't need a new story. The parent is not going to repress what they see because it's too depressing or too tough to take in. The parent is going to summon superpowers and focus all of their being on rescuing their child.

We're approaching an invisible boundary beyond which there is no return. What we are doing now, very soon will be impossible to undo. And those precious few who have realized the seriousness of our situation are trying to warn the rest of us.

But we are social animals, we follow the herd, and as long as our leaders are not behaving as if we are in crisis, who is going to understand that we in fact are? We're waiting for the leaders of the herd to signal us all to halt. And then lead us out of danger to safety.

Scene 76.
Performance Review

'We don't really know what to do with you. You're a bit of a hopeless case,' the teachers at the University College of Opera said when it was time to do the final production.

The rector said the same thing. 'We don't really know what to make of you or where you fit in.'

They didn't say it in a joking or positive way, but as if I'd done something wrong.

Something about me irked them.

I was doing two courses in parallel: one at the Swedish Royal College of Music and the other at the University College of Opera. At the same time I was singing in the Radio Choir under Gustaf Sjökvist, and working at Oscarsteatern as a dancer and understudy for the female lead in *Cyrano de Bergerac*.

My student grant was used up, I had to provide for myself, which I had no problem doing. I was having a great time. Jumping from one thing to another suited my restless nature, and what an education it was.

I sang and slept and danced, and I had time for everything. Except possibly socializing with the opera class and going to all their parties, where I still felt just as awkward as I'd always felt at parties. But it was okay, I didn't mind, because not only had I finally found a place where my way of doing things worked – it was working very well.

Until the day when I walked into a stage performance class at the University College of Opera.

'This has nothing to do with me,' said Philippa, our teacher, 'but the other students have demanded to have a meeting during this class. A crisis meeting.'

And then I was asked to sit down facing a large semi-circle of chairs. It was troubling that I wasn't part of the group, I was told. That I'd missed so many of the college's parties.

I had committed the crime of not belonging. They told me that I thought I was somebody, but evidently I was no one. Because according to them I was pretty worthless, actually.

And the punishment for choosing to go my own way was that I was now forced to learn a lot of things that I'd never heard before. Such as that it's not okay to be different, because if you are it will work against you. This insight became a continuing source of sadness for me.

I withdrew, locked myself in my apartment, and there, four flights up on Kungsklippan, I devised a way of making the worry and anxiety disappear. All I needed to do was to binge eat and stick my fingers down my throat. Then I'd feel great again. All I had to do was vomit and that lump in my stomach would disappear, and sometimes it wouldn't come back for days.

Bulimia is a very dangerous disease, so it was not a sustainable solution, but it was the only thing that soothed me at the time.

The problem with it was that you can't sing when you've vomited.

And it was a helpful problem to have, because I can't live without singing, and suddenly I was forced to choose.

I chose song.

And song saved my life.

Scene 77. Svenny Kopp

'The basic problem with ADHD is that you're operating by the pleasure principle. You can only do things that you're absolutely interested in. Otherwise you won't do them. Because it's to do with the reward system. It's to do with dopamine levels.'

– Svenny Kopp, University of Gothenburg

One day in early May 2017 I go to a lecture by chief physician Svenny Kopp. She's a researcher and neuropsychiatrist, and considered to be an international pioneer in paediatric and adolescent psychiatry. This is because her research focuses on something as original as girls.

The audience consists of my friend Gabriella and me and several hundred people from the Stockholm school health, paediatric and adolescent psychiatry community. The children's psychology industry people, in other words.

Gabriella is like me, and that is probably why she's the only one I can bear to spend time with right now. She has a daughter with a diagnosis and is constantly teetering on the brink of a breakdown.

By all accounts, she's long overdue for one – but she is strong and fights on, like so many others I know in similar situations. Only the truly strong ones crash, because they're the only ones who manage to push themselves so far beyond all reasonable limits that at last they fall apart and burn-out: this relentlessly demanding discipline in which women's contributions are in every way superior to men's.

What Svenny Kopp has noticed in her research and her clinical experience is that speaking in general terms of 'children and adolescents' very seldom benefits girls.

Kopp begins her lecture. 'We must unfortunately – or perhaps this is a good thing – divide patients up into girls and boys. Teenage girls, teenage boys. Because they do in fact live under different conditions. [ADHD] manifests in different ways. And if we talk about "children and young people", we usually mean boys.' She speaks in a broad Gothenburg dialect and is nothing like anyone else we've listened to before. She is speaking directly to Gabriella and me. She is telling it like it is.

'Too few girls are diagnosed with ADHD and autism. I still get completely obvious cases, where I think, how is this possible? How can anyone call this a "teenage problem" or a "family relationship disorder" when it is so obviously an ADHD problem?'

The fact that a woman researcher is illuminating the structural inequality within paediatric and adolescent psychiatry is clearly provocative. After a while several members of the audience get up and leave. Others groan and sigh, and I think about something I saw on my phone: 'When you're used to privilege, equality feels like oppression.'

For Gabriella and me it's like seeing our favourite band.

'I'm almost a little star-struck,' Gabriella says, and I can only agree. Especially when Kopp says how girls are so obviously disadvantaged in relation to boys, who are heard and seen and early on take possession of the few resources that are still left for teaching support and special education.

Svenny Kopp continues: 'This means that the boys end up getting support much earlier. And then we discover the girls later on, in their teens, and on the one hand they don't want any support at that age, they want to be like everyone else. And on the other hand, by this point they're competing for what resources are left over, which is much harder. It's here that parents need to be equipped with tools too.'

Kopp takes a sip from the glass on the podium.

'How should you cope with a girl who doesn't get up in the morning? Can you drag a fourteen-year-old girl out of bed and carry her to school? Of course you can't, right? What should you do? What should you do when she doesn't do her homework? How will you manage all these conflicts, the irritability? This tiptoeing around? This not being able to maintain order? How will you manage all these everyday situations? These are not easy things.'

At the break I show Gabriella an article I've just read. It's about a research study on *children* with ADHD, and the study was done on sixty-four children, *and all sixty-four children were boys*. The mere thought that scientific investigations on 'children' in 2018 can be done without equal gender distribution really says it all.

It's not easy to get a neuropsychiatric diagnosis, despite whatever you may have heard or the opinion pieces you may have read.

It's difficult. Especially for girls. Because how can a girl fit into templates and criteria designed for boys? Girls couldn't even have Asperger's or ADHD a few years ago.

More or less everything regarding diagnoses is still based on boys. The basis for assessment, medication and information.

Of boys, for boys, about boys.

*

All diagnoses look different from individual to individual, and in girls they can express themselves quite differently than in boys. One example is that boys with ADHD often act out, whereas girls will often behave in exactly the opposite way.

The majority of diagnoses are as a result of behaviour that is understood as disruptive to others, and because girls often bottle things up inside, they immediately fall behind. Those not seen or heard outside the home seldom get help.

Because how many parents have the energy to get to the bottom of all these problems? How many parents choose to devote three or four years to wage war on paediatric and adolescent psychiatry so that, in the best case, their child will get a stamp on the forehead labelling them, in many people's eyes, as handicapped?

Many parents now know all this because the studies and findings are available to read on the internet. But many *within* adolescent psychiatry choose not to acknowledge these new findings – the research moves quickly, but the criteria and practice don't always keep up. And a great many children end up in that gap. Primarily girls. Girls who often get stuck in long-term involuntary absence from school which can mean the start of life-long alienation. Girls on their way into the risk zone created by undiagnosed functional disorders such as Asperger's and ADHD; a risk zone bordered by eating disorders, OCD and various self-harming behaviours.

Countless studies from around the world show that ADHD is associated with a greatly elevated risk of addiction and criminality. Although the correlation between ADHD and eating disorders is a brand new research area, clear signs of strong connections are already being seen.

*

When girls at last – finally – started getting diagnoses too, suddenly large parts of society started shouting *there's an inflation in diagnoses!* And that is as crazy as it sounds. But this ignorance is not the boys' fault. Their problems obviously aren't lessened because girls have been ignored, and they need all the support they can get. Besides, the boys – and their parents – have to put up with the negative effects of the ADHD spotlight, such as their characteristic 'behaviours' being openly mocked, even by school staff and parents. Such as everyone finding the solution in constantly recurring and immoderately popular opinion pieces with titles like 'Sit still and learn some manners!'

Even though the research says something quite different.

Up on stage, Svenny Kopp is approaching the end of her lecture.

'What I have done during my years as a researcher is to study both girls with autism, ADHD and Tourette's, and girls without any diagnoses at all, and I can say that these are families who live on different planets. It's almost impossible to imagine how great the difference is, and what pressure these families live under that have girls – and boys too, for that matter – with some sort of diagnosis. It is extremely stressful.'

It is silent in the hall, but for a few cautious coughs and someone turning the pages in their notepad.

'The divorce rate is higher than in other families. And primarily it is the mothers whose stress levels are almost . . . yes, it's not tolerable nowadays . . . it's not. We can't take care of these complex family problems, we don't live in "Welfare Sweden" any more. We cannot deal with the stress that primarily mothers are subjected to for years, while often they are met with a lack of understanding from officials.'

*

On the way out of the auditorium, Gabriella tells me about ten-year-old children who are suffering from depression, chronic fatigue and burn-out syndrome, and it sounds like a bad Social Darwinist joke, but I know it's true. I've seen it with my own eyes. She tells me about a girl with Asperger's who hasn't got out of bed for two years and who can no longer walk because her Achilles tendons have contracted so much.

And I think, who has the energy to give them a voice?

Who can shout loudly enough so that everyone stops and listens? No one can. Not on their own.

Scene 78.
Welfare State Child Deluxe

I grew up in a small industrial town in the 1970s and I lacked for nothing. I was a Swedish welfare-state child deluxe. When I look at the children who are growing up today, thirty-five years later – when I see my own daughters – I think that I wouldn't have stood a chance.

The speed, the volume, the intensity, the demands for profitability and results that permeate everything. The arts school that abandons individual lessons in favour of cost-effective group instruction that once again excludes any child who lacks the capacity to function in a group.

All those whose difference could be transformed into creativity, self-confidence and art, but who now risk disappearing into yet more exclusion and alienation.

Yet another financially profitable failure.

Scene 79.
Next Door to Seinfeld

In some shabby offices above the Manhattan diner that would later be immortalized in the TV series *Seinfeld*, a new section of NASA was established in the 1960s. This institute was working on something called 'the greenhouse effect'.

It has now been over thirty years since its director, James Hansen, testified before the United States Congress and presented evidence confirming that global warming was real.

On 24 June 1988, the *New York Times* reported on Hansen's testimony: 'It was 99 per cent certain that the warming trend was not a natural variation but was caused by a build-up of carbon dioxide and other artificial gases in the atmosphere.'

But who outside the climate and environmental movement has even heard of him? And how many of us know the results and meaning of the research that he and countless others have continued to do in this field?

If we took the climate issue seriously, Hansen would be world famous and each and every Nobel Prize would in some way or other have a connection to the sustainability crisis.

But that's not how it is.

James Hansen's prognoses proved right with an unpleasant clarity and yet he is still something of an outcast, overlooked and counteracted by every contemporary US president. He is also a prominent critic of the, in his view, hopelessly insufficient Paris Agreement.

'Promises like Paris don't mean much, it's wishful thinking? The real hoax is by leaders claiming to take action. It's a hoax that governments have played on us since the 1990s,' says the former NASA director, who is now professor emeritus at Columbia University.

He has a point.

Because in the thirty years that have passed since his testimony, the world's CO_2 emissions have not decreased one bit. On the contrary, they have gone up by 68 per cent, and despite all the renewable energy – all the newly installed solar and wind power – the world uses more fossil fuels today than in 1988. We are still moving in the wrong direction.

Scene 80. Superpowers

When the social media #MeToo campaign started hammering on the surface that feminists have been scratching at for half an eternity, an opening came to light.

A crack in the facade.

Through that opening voices that had sounded for decades suddenly started to be heard, and when you least suspected it, suddenly a small miracle occurred.

Although the miracle was no miracle.

It was only a few joint editorial decisions.

Because when the media chooses to push an issue, as in Sweden with #MeToo, everything changes.

Many within the environmental movement are hoping for a similar breakthrough for the climate.

Although without the scratching on the surface, and 130 years of tiny, tiny steps forward, of course.

We just don't have that kind of time.

In fact, we have no time at all. 2020 is the year that a revolutionary adjustment needs to already be in full swing.

'Insight is lacking about the radical changes that are required,' says Professor Johan Rockström, previously head of the Stockholm Resilience Centre and now director of The Potsdam Institute for Climate Impact research.

We are in a crisis that has never been treated as a crisis.

The reports are many but the reporting is minimal. According to the Swedish pollsters SIFO, the environment had the lowest visibility among political issues in the media in 2016, at the same time as the annual Society–Opinion–Media (SOM) report from the University of Gothenburg once again noted that 'changes in the climate' were what worried us most.

The media's handling of the climate and sustainability issue is nothing less than a total failure. The question of the fate of humanity is – in the best case – reduced to scattered articles, news items or feature sections in which climate news is isolated, while paper editions and news websites are bulging with travel reporting, shopping tips and car articles.

On the radio and TV, debate is what matters, where it's word against word.

There are no headlines. There are no extra news broadcasts. There are no emergency meetings. There are no public information campaigns.

It's 'economy before ecology', and for that reason the crisis should not be treated as a crisis but instead as an opportunity for new 'green' growth. *That* is what the plan to save the world looks like: a strategy that teaches that alarming reports which could give people insight also risk making those same people collapse into one big '*Oh no! So the climate crisis was for real? I had no idea – but now that I know I'm giving up. Because if the Paris Agreement means limitations for me personally then I prefer a full-scale Venus effect, with sea levels rising sixty-five metres, mass deaths, extinction, and a violet-coloured, acid-bubbling ocean!*'

That's the lie of the land.

News editors must neither frighten, blame nor tell it like it is, because then the rest of us might stop doing all that super-duper climate work that we've already started – you know, all that stuff that makes the CO_2 content in the atmosphere keep increasing ten times faster than during the greatest mass extinction that ever took place. Oh heavens no, instead we're supposed to tell a new story – a positive narrative. Something it's possible to 'like' on Facebook.

But you know what?

There already is a new story. And it is so positive that the angels are singing hallelujah and doing somersaults across the sky – because we've already solved the climate crisis and we know the solutions are going to work.

The solutions are so brilliant that they'll solve a great number of other problems all at once, such as growing wealth gaps, mental illness and inequality between the sexes.

With the proviso that these solutions require fundamental changes and a concession or two in return.

Such as a very high CO_2 tax.

Such as our overall goal having to be the reduction of emissions.

And to start planting enormous quantities of trees while allowing most of the existing forests to remain so that they hold on to the CO_2 they've already absorbed. Forests are our salvation. But we must start treating them with the respect they deserve.

The solutions require that we slow down and start living on a smaller scale, collectively and locally. That means everything from local democracy to more collectively owned energy and food production.

That we cooperate, because collective problems require collective solutions.

And that instead of spending over 4 trillion kronor to subsidize fossil fuels every year, the world invests that money in building wind and solar plants. A sum that we can surely multiply.

We can, if we want to.

But not without concessions on our part.

Such as investing in existing technology instead of waiting for things that might come later, once it's already too late.

Such as changing many of our habits and many of us having to take a few ecological steps back.

And the companies that have created the problems paying for everything they have messed up – companies that, even though they knew the risks, have seen unbelievable earnings as a result of destroying the climate and our ecosystems.

We're not the ones who messed things up. It's not everyone's fault. On the other hand it is our joint responsibility to secure the planetary conditions for future generations. Their future is in our hands.

If you're among those who believe that technology will save us, then I recommend you place yourself at the very top of an Olympic ski jump and look down over the target area. That's how steep the emissions curve looks – the curve that, starting today, we must get down to net zero. The graphic curve that ought to be displayed on every front page of every newspaper in every country.

Our fate rests in the media's hands. No one else has the reach required in the time we have left to act.

And we can't solve a crisis situation unless we treat it as a crisis situation.

Everyone who has ever witnessed an accident knows what I mean.

In a crisis we get superpowers. We lift cars, fight world wars and climb in and out of burning houses. It only takes someone falling down on the pavement for a line of people to appear, prepared to drop everything to help out.

It's the crisis itself that is the solution to the crisis.

Because in a crisis we change our habits and our behaviour.

In a crisis we are capable of anything.

The vast majority of us are going to feel much better once we slow down and live more locally, safe in the knowledge that our children will get the chance to develop the inventions and solutions that we haven't managed to invent ourselves.

The vast majority of us are going to feel much better when entire countries are given a chance to live instead of us forever being on our way to the next big city, the next trip, the next airport, the next whatever.

The world gets bigger the slower we travel.

And *everyone* will feel much better in a society that puts sustainability first.

Scene 81.
Empty Words

The election campaign has started.

It's July 2018 and suddenly all the politicians are talking about the climate crisis. It can't really be avoided any more. Because after months of unprecedented drought and heat, that thing the experts have warned about for decades is actually happening. The harvests are ruined, the ground water is running out.

Sweden is burning, in forest and peat and bog, from Gällivare and Jokkmokk in the far north down to meadows in the south. And after only a few days of crisis reporting in the media, at some level we start to understand the fact that almost one-sixth of the country is north of the Arctic Circle – in the Arctic, where climate change is expected to be the most drastic. The climate crisis is no longer somewhere far away – with 1°C of warming it is already here and we find ourselves on the front line.

But it's painfully obvious this isn't something our elected officials want to talk about. And almost none of them say a single word about either cause or consequence. Politicians – needless to say – are there to win elections. And you don't win elections by telling it as it is – you win elections by telling people what they want to hear.

'The climate is the critical issue of our time,' everyone starts to say in passing, while presenting an analysis as deep as a tabloid horoscope. It's up to other countries to do everything – anything else is 'bumper-sticker politics'.

No one mentions that over half of Swedish emissions aren't even covered by the statistics. It's more comfortable that way.

No one talks about the fact that Sweden's ecological footprint is among the highest in the world. No one breathes a word about the fact that, since international flights aren't counted by official statistics, a bus trip between Sandviken and Gävle produces higher emissions than a round trip to New Zealand in business class. Just like ocean freight or goods we import from other countries.

When we moved production to low-wage countries we didn't just get out of paying reasonable wages – we also got rid of a gigantic part of our CO_2 emissions. So now we can claim that everyone else should act, because we've already improved things by dumping our factories in China, Vietnam and India.

'They need our industries and our trade to raise their standard of living,' you can obviously argue. But already, at 1.5°C of warming, it seems there are other things they need more. A habitable environment, for example.

'Sweden is too small,' say the politicians who represent the majority of the voters. 'It's better that we try to influence others.'

And no journalists counter that rhetoric. No one says that by the same logic we could all refrain from paying taxes because 'My little contribution is so ridiculously small in the big picture that it's better that I don't bother with it and invest in things that really benefit me and my family instead. Anything else is virtue signalling.'

No journalists mention that when small countries like Costa Rica decide to prohibit single-use plastics, that generates news articles that are shared hundreds of thousands of times, because the world is so starved of positive examples. Starved of what people understand as hope. Hope – as in prohibitions and limitations for the good of everyone. No one mentions that little Costa Rica started a trend and that other countries have already chosen to follow. Countries like India, for example.

There are Swedish politicians with insight, who want to tackle the problems, but they aren't heard. Public opinion is too weak. The debate has never started properly and the gaps in understanding are too great. Some find themselves at point 117 while the majority haven't even got to point 2.

We're reading *Factfulness* by the three Roslings, in many ways a fantastic book, but not even there does the climate and sustainability crisis stand out as particularly acute.

> Those who care about climate change should stop scaring people with unlikely scenarios. Most people already know about and acknowledge the problem. Insisting on it is like kicking at an open door. It's time to move on from talking talking talking. Let's instead use that energy to solve the problem by taking action: action driven not by fear and urgency but by data and cool-headed analysis.

So write three of the world's foremost – and rightly acclaimed – popularizers of our time. But this train of thought is far from unique to the Roslings' Gapminder Foundation.

Really, it could come from any editorial board, politician, decision-maker or business representative. This is mainstream. This is the picture that counts.

But is it correct?

Is the information spread by environmental organizations and climate experts improbable? Are thousands upon thousands of researchers out to frighten us?

And above all: do we have time to continue, without urgency, with yet more cool-headed analysis?

Or is it the case that the changes are now coming so fast

that we don't even have time to register the information? It's always that little detail about the concentration of CO_2 in the atmosphere that upends everything for every single one of us.

Almost nowhere in our culture is it said that the climate issue is an indication of a system error. It's a *problem* and you solve problems with new inventions. New gadgets. And when the research says something else you order new studies and new researchers whose opinions might better align with what we want to hear, and then we simply carry on like that. Round after round.

It is a life-threatening development, but to me there is one thing that chafes a little more than all the rest, and that is the constantly repeated assertion that 'Most people already know about and acknowledge the problem.' Everyone seems to think this is the case.

'When we look our grandchildren in the eyes in the autumn of our lives we can tell them that we managed the threat when we saw it. Or else we can tell them that we did nothing, even though we knew,' says Sweden's deputy prime minister when she concludes the annual Almedalen gathering of Swedish politicians in 2018.

No one seems to question that 'we all know', even though it basically involves a view of humanity that is utterly foreign to us all.

Because if we knew – if we really understood the consequences of what we're doing and kept on doing it . . . what would that say about us?

Scene 82. Differentness

I'm pretty hopeless actually. I can hardly manage any practical things at all.

I don't have a driving licence.

When I was twenty I heated up bread still wrapped in plastic in the oven; and I've never been able to log on to internet banking to pay my bills.

I have to write long to-do lists, because there's no other way I'll get it done. I can't let certain things go. I get stuck. If I hadn't been a singer, to be honest I probably wouldn't have been anything. In all likelihood I would have got stuck in one of those abysmal holes where people with undiagnosed ADHD sometimes end up.

Today extroversion is essential. You have to know a little about a lot. In principle you can possess research-level knowledge and still not fulfil the grade requirements in high school if you fail to assert yourself verbally.

So what happens to those of us who are very good at something specific but who lack the capacity to do anything other than what interests us?

What happens to those who simply happen to be a little shy? What happens to those who feel physically ill from talking in front of others? What happens to the large portion of the population that lacks the social competence we value today above all else?

Would anyone who deviates too much from the norm survive school in Sweden today? I doubt it. I believe that many of those who one day will be active in areas that demand sensibility, listening and empathy would not. So we have to change that.

Too much is at stake.

Differentness is the basis of all art. And without art everything is going to slowly, slowly crumble into nothing.

Scene 83.
Behind the Scenes

Greta, Svante and I meet Kevin Anderson together with his research colleague Isak Stoddard at the Centre for Environment and Development Studies, Uppsala University.

We've been in contact with Kevin and Isak before. In June 2017, with Björn Ferry, Heidi Andersson, Staffan Lindberg, the meteorologist Martin Hedberg and others, we wrote an opinion piece in *Dagens Nyheter* that attracted a lot of attention, in which we explained why we chose to stop flying and stay on the ground. To a large extent this article laid the groundwork for the debate on flying that took off in the media a few months later.

It hasn't rained for several weeks and the lawns in Uppsala are already brown, baked by the late spring sun. Kevin Anderson tells us about the heat in his guest apartment and how he sleeps with his window wide open 'like in Greece'.

We fill our cups with fresh coffee and oat milk and sit down in a little conference room with couches and bookshelves. Kevin drinks tea.

'First and foremost,' says Svante, pressing the record button on his phone, 'when you talk about how much countries like Sweden have to reduce carbon emissions, the figures vary. You and other scientists say 10–15 per cent per year, but politicians and the Environmental Protection Agency are talking about 5–8 per cent. How do you explain that?'

'There are several reasons. The 5–8 per cent figure does not include, for example, aviation, shipping and goods produced in other countries,' Kevin begins. He talks fast and clearly, with a persuasive conviction like few others we've ever met.

'Then the calculations from the industrialized countries, like Sweden, never include anything of the fairness aspect for countries in poorer parts of the world that we have undertaken to address. It states [this] clearly in the Paris Agreement, the Kyoto Protocol and so on. We have committed to reducing our emissions based on a perspective that we completely ignore.

'But most important of all: our emissions-reduction models are highly dependent on enormous quantities of negative-emissions technologies sucking CO_2 from the atmosphere at some point in the future. That is, reliance on inventions that don't exist at scale and that many researchers say will not work at the levels assumed in virtually all climate models. Until very recently, few climate researchers understood the huge scale of reliance on "negative emissions". I know many colleagues who were shocked when they realized that such highly speculative technologies are not just included in a few future calculations, but in almost every single one.'

Kevin pauses suddenly and repeats his research colleagues' reaction. I mostly sit silently and let Svante keep track of our notes and questions. As always when we talk to climate people I prefer to listen. Partly because I learn more that way, but mostly because I'm afraid of making a fool of myself by saying something stupid.

'Isak and I have calculated that for 2°C, rich countries like Sweden must start reducing their emissions by 10–15 per cent per year, starting today, so that by 2025 we will have lowered emissions by as much as 75 per cent. That means all our energy must be zero emissions by 2035–2040, including flying, shipping and all transport – they must all be zero carbon!'

All eyes in the room meet for a brief moment. We find

ourselves very far from what we usually read and hear about the so-called 'green transition' that politicians and the business community so happily talk about.

'According to our calculations [Sweden and the UK] have under ten years left at the current rate of emissions. And we haven't even included goods manufactured in other countries. If we were to include those we have even less time left,' Kevin continues. 'Sometimes I like to end my lectures by quoting the American futurist and sustainability author Alex Steffen, who says about climate change: "Winning slowly is the same thing as losing." He's quite right, because we simply don't have any more time: the transition and the action must start now.'

Sweden recently passed a climate law that many decision-makers are extremely happy about and proud of. And even if the thought of such legislation is something positive, neither Isak nor Kevin are particularly impressed.

'The Swedish climate law must immediately be rethought if it's going to have any progressive effect,' says Kevin. 'It must first and foremost include a CO_2 budget and a fairness perspective in line with the Paris Agreement, making allowance for countries that have yet to build their welfare system and physical infrastructure. Such allowances need to be at the core of our calculations and our climate laws. Then obviously we must also include emissions from our international shipping and aviation sectors.'

When Kevin lectured at the Royal Swedish Academy of Sciences in September 2017, with Crown Princess Victoria present, he started by issuing a 'health warning' to the audience, because what he does research on is not easy for the majority to absorb. A couple of years ago we probably would have needed something similar prior to this conversation with Kevin and Isak, but now it has all become part of our everyday existence.

'For almost thirty years we have known all we need to know about climate change, but during all that time we have chosen not to do anything about the problems. Not even progressive countries like Sweden have done anything; if you include aviation, shipping and goods that are produced abroad, Sweden's emissions remain at the same level as in 1992 when the UN's first climate conference was held in Rio. Instead, we've let economists control our decisions. We fool everyone into thinking that we're doing what we need to, but the fact is that no industrialized country in the world is doing anything that comes even close to what is required. There is a perfect Swedish word for what we are doing: *swindlee*.'

'*Svindleri*,' Isak corrects him. Swindle.

'That's right! *Svindleri*!' Kevin laughs and continues.

'If we had acted then, like we said we would, the climate issue would not have been such a big problem. We could have managed everything with new technology and changed economic guidelines. But because we have now spent thirty years talking, lying and delaying, the system has to change completely, because today's economic model is not going to be able to solve the climate crisis. Much less the sustainability crisis. It has to be replaced.' Kevin shifts on the couch, which, suitably enough, appears to come from a flea market, like the rest of the worn, mismatched furniture.

'But there's a lot that is hopeful right now. Many indications that a system change is possible, and even if the results of much of the change that has taken place aren't always great, the signs are there. The financial crisis, the Arab Spring, Corbyn, Trump, Bernie Sanders, the price of renewable energy, the debate on the effect of diesel and petrol on our health.'

'And the MeToo movement,' I interject.

'Exactly,' says Kevin. 'We are probably facing major changes in society. That's extremely hopeful.'

I lean over to Greta and ask if it's okay if I tell them about what she is planning to do. She nods.

'Greta is thinking about going on school strike in front of Parliament when school starts in August. She intends to sit there every day until the parliamentary election.'

Kevin and Isak's faces light up and they stop talking.

'So how long will you be sitting there?' Kevin asks.

'Three weeks,' Greta says, almost inaudibly.

'Three weeks, you say?' Isak checks.

Greta's face shows agreement.

'That ought to get a few politicians to listen,' Kevin says approvingly.

'She got the idea when she was part of a telephone meeting about starting a Swedish version of Zero Hour. It's a new movement in the USA that will get children to take politicians to task for not doing anything,' Svante says. 'But Greta doesn't think protest is enough any more. She thinks some form of civil disobedience is required. Something agreeably illegal. Is that right, Greta?'

Svante checks, as usual when he is speaking in Greta's place, due to her mutism. She nods.

'But in this case she'll have to do everything by herself. We can't stand behind and help her,' Svante says.

'But Greta already knows the issue better than either Svante or I do,' I interject. 'It's only because of our daughters that we even opened our eyes to the climate crisis. Without them we never would have got involved.'

'Well done, Greta,' Kevin and Isak say in unison. Greta's eyes glisten and I sense that something is starting right there and then, in the feeling of being seen and heard in a context that has real significance.

There is a pause. These thoughts take over the room – that the almost invisible little girl on the chair by the window is planning to put herself at the very centre of the spotlight, and, all alone, in her own thoughts and words, question the foundation of the prevailing world order . . .

To find a path that can lead us out of the sustainability crisis is to do the impossible, and I love everyone who is crazy enough to try. On the other hand, when it concerns my child I am far more circumspect, and if it were up to me I'd probably say no. But it's a long time until August. And it's a long way from thought to action when you haven't even finished eighth grade.

A few years ago Kevin attracted some attention when he refused to take part in a climate conference in London because all the participants were forced to pay a fee to offset their emissions. Kevin argues that from a total emissions perspective offsetting will almost certainly do more harm than good because the signal it sends is that there is an easy way to recapture the CO_2 we release, and thereby undo the emissions.

'So if the idea of offsetting emissions is wrong,' Svante asks, 'is there no other way to compensate for what we're doing other than refraining from actually doing it?'

'No. First, if you fly then you're sending a clear market message to the airlines that they should keep doing what they're doing and buy more planes and expand airports. Which is exactly what's happening all over the world right now. Airlines are ordering more and more new aircraft and airports are expanding. By choosing to fly, you're not putting pressure on politicians to invest in trains instead, which they ought to do. And second, you are sending a lot of CO_2 into the atmosphere that's going to affect the climate for thousands of years. It's not going to disappear just because you buy solar panels for poor villages in India. That isn't to say that we shouldn't buy solar panels for poor villages in India, or plant trees where it does ecological good to plant trees – which is not just anywhere, because soil, reforestation and emissions are extremely complicated matters about which we still know too little. But we don't need to fly or eat hamburgers to justify spending money on worthwhile projects. Carbon offsetting is like paying poor people to diet for us.'

Scene 84. When the Microphone is Off

After several hours at the university, we walk over to the Botanical Gardens and have lunch in a café next to the Tropical Conservatory. Roxy finally gets a big bowl of water, which she gulps down in the heat before slinking under the table and settling down. We order a vegan lunch. Greta saves her glass jar with bean pasta for the trip home.

'I often eat the same thing every day too,' Kevin says to Greta. 'I mostly live on broccoli and bread. Everyone thinks I'm joking when I say that, but it's simple and practical. And I do like broccoli and bread a lot.'

Greta nods subtly in response and I assume that he's saying this half-jokingly and more out of empathy than eccentricity.

We talk about our summers in Lewes when Greta was little, and of course I talk about my childhood memories of the convent in Whitby. We step into the role of English-speaking Swedes and as usual become slightly different people than we are in Swedish.

'You must come to Dalhalla,' we say, and Kevin wonders how far it is from Uppsala. 'Two hundred, maybe two hundred and fifty kilometres?'

'You can cycle,' Isak says, in a way that makes it clear that Kevin is an extremely good cyclist.

Kevin is like any pleasant, sociable Englishman. Funny, open and sympathetic.

We talk about friendships being put to the test when you let the climate crisis influence and change your life, but Kevin says that he has never had any major problems with that.

'I never get angry at climate-change deniers or sceptics,' Kevin says. 'Not even politicians or decision-makers irritate me all that much. The only thing that truly upsets me is other scientists who more or less consciously distort their scientific analysis and conclusions so that they will appear less alarming than they are. That makes me angry.'

When you see Kevin talk on stage you understand that on some level he is angry, but he never *looks* angry. More like impassioned, factual and full of conviction. When you listen to his voice you can definitely hear traces of fury, but he never sounds upset.

'There are those,' Kevin continues, 'who think that we scientists shouldn't tell it like it is because it's too political. I think it's the other way around. It's those who choose to be silent who are truly political, because their silence says that everything is okay, and that's an extremely powerful message that supports the status quo, or business-as-usual. Many researchers also say that our message can't be dealt with within the current political and economic systems, and for that reason we have to adapt what we say to the prevailing reality. But once again I think that's wrong. We climate scientists are only climate scientists after all. Our mission is to present facts about the climate. We aren't experts in politics or social issues, so it isn't our place to let the politics – or the thought of how our results are going to be received – guide our work. Our mission is to do our research carefully and communicate our conclusions clearly and bluntly.'

We finish our lunch and put the trays away. Eager to snap up three abandoned scraps of bread, Roxy overturns the water bowl. In order to explain many of his scientist colleagues' aversion to plain speech, Kevin returns yet again to the years after James Hansen's testimony to US Congress and the UN's first climate conference in Rio de Janeiro.

'When we started working with this issue after Rio in 1992, there was great optimism about how we could solve the problems, and that positive spirit lives on today. Such optimism was legitimate then. But, as the years have passed and nothing meaningful has happened, the problems have of course accumulated. Yet we still have that optimism we felt in the beginning. Many scientists have been a little like the frog in the pot. At what point should we jump out?'

'But has that started to change?' Svante asks.

'Yes. Now, when climate impacts are happening so much faster than any of us expected, we see more and more researchers choosing to be more outspoken. But it happens very gradually and they still usually choose to soften and tone down the message when they speak publicly. If, for example, you have a beer with a scientist or an informed politician, they'll discuss how bad the situation is. But put a microphone in front of them, and all too often they'll trot out some optimistic nonsense about climate change.'

We leave the shade of the apple trees and go out into the baking Uppsala sun. On our way back to the university we ask Kevin and Isak how often they are asked to take part in media and public broadcasting.

We know how unapproachable Sveriges Television is where programmes about the climate are concerned, but we want to ask anyway – if for no other reason than to have something about this on record from 2018. We've tried to present and offer various programme ideas ourselves, but not even with Sweden's most well-known and successful TV producer attached was there any interest.

'We listened to Johan Rockström's summer talk and it left us with a sense of hope. As if we're going to fix this,' a programme commissioner said when they declined six segments of infotainment about climate and sustainability.

But for a long time now one of the world's leading climate researchers has been active in Uppsala for several months a year. SVT or TV4 must have used this opportunity to do something on this subject that causes Swedes the most worry . . . ?

But no. No inquiry from national TV has come in. 'On the other hand we have seen a strong increase in interest from the media since Kevin arrived,' Isak explains. 'We often get inquiries and we've contributed quite a bit to radio, newspapers and regional TV.'

'Okay, but how many times has Kevin appeared on *Rapport*, *Aktuellt* or the TV4 news?'

'Never,' Isak answers.

'How many times has Kevin been *asked* to appear on *Rapport*, *Aktuellt* or the TV4 news?'

'Never,' Isak repeats.

'Been interviewed in *Dagens Nyheter* or *Svenska Dagbladet*?'

Isak makes a face and once again shakes his head before he repeats the answer for the third time.

'Never.'

On the ride home to Stockholm in the electric car we think about all the things we forgot to ask.

But it doesn't matter, because those questions never needed to be asked.

You see, there is no resignation, no darkness or gloom in Kevin Anderson's vicinity.

There is simply the calm, hopeful and solid energy of taking the initiative.

Scene 85. 'It's Never Too Late to Do as Much as Possible'

– Pär Holmgren

If the history of the earth were to be translated into one year, the Industrial Revolution would take place roughly one and a half seconds before midnight. On New Year's Eve.

During this – in historical terms – infinitesimally brief period, we have already caused so much destruction that our progress can only be compared with the earth's previous five mass extinctions. Although with one major difference: time.

Events that without human involvement would have taken hundreds of thousands or millions of years we can knock off in a few weeks, just by living as usual.

We are living in what is often referred to as the sixth mass extinction. And it did not start at the end of the eighteenth century: it's been going on for thousands of years.

Many people think there was a time when humans lived in harmony with nature, but no such time has ever existed.

People have lived in harmony with nature. But never humankind.

Wherever we have ventured, extinction has followed in our tracks. The connection between humankind's appearance and the number of extinct species – primarily really large animals known as 'megafauna' – speaks for itself.

But it is a language that we can relinquish. If we want to.

Scene 86. Testament from an Era of Historic Abundance to All Future Generations

There will come a time when we are no longer here.

There will come a time when our children and grandchildren and their children are no longer here.

A time when, in the best case, we live on in a family tree, a hard drive or some dusty photograph where no one recognizes anyone any longer.

Sooner or later we are all going to slip into oblivion, however important, hated or loved we might have been during our lifetimes.

It's a difficult thought. And it doesn't get any easier once you realize that the things we will leave behind in the end will be quite a bit more than just some memories, deeds and photographs. Because it turns out that the sound, humanistic upbringing most of have been granted omitted one small detail: our ecological footprint.

You see, there will be a time when we are all gone and forgotten, and the only thing left of us will be those greenhouse gases that we more or less unwittingly sent up into the atmosphere.

On the way to work.

At the supermarket.

At the shopping centre.

Or on the way to a TV recording in Tokyo.

Some of them are going to hover around up there for a thousand years.

Some will be bound up in trees and plants.

And some may be vacuumed up and stored deep down in

the bedrock by means of some invention and some smart logistics that haven't been created yet.

Perhaps a magic vacuum cleaner will be invented for the oceans too; a machine that can clean our oceans of all the CO_2 that has been absorbed there. It will no doubt be needed, because up to 40 per cent of our CO_2 emissions are soaked up by the oceans and cause acidification that many consider to be an even greater threat than the greenhouse effect taking place in the atmospheric layers above the surface.

So we will all live on for centuries – but perhaps not in the way we've imagined.

Because, with very few exceptions, the memory of our existence is not going to survive the ecological traces we've already left behind.

If these thoughts feel a little dark and hopeless right now, keep in mind that all it may take is a single big pop star or influencer to start redrawing this map. The power of celebrities is by all means controversial, but this is the reality we live in today and we don't have time to change it. The advantage is that, in today's connected world, it's enough for just one king, superstar or pope to choose to strive towards zero emissions – with all the veganism, rejection of flying and solar panels on the roof that entails – for change to feel more possible.

No one can create a change in our whole system on their own. But one single voice could be enough to start a chain reaction that can set everything in motion – if that one voice is powerful enough.

IV

Imagine if Life is for Real and Everything We Do Means Something

'But man is a part of nature, and his war against nature is inevitably a war against himself.'

– Rachel Carson, *Silent Spring*

Scene 87.
Further North, July 2018

It's hot in Luleå. Really hot. Svante wipes the sweat from his brow, flaps his shirt to air it and exhales loudly and demonstratively. But the woman at the reception desk wants no part of that kind of wordless comment.

'Now that it's finally really warm up here I don't want to hear anyone complaining,' she says, as if this was about a topic of greater importance than some old weather.

'Of course,' he replies, entering the code on the card reader. It's important to pick your battles.

Greta and Roxy are waiting on the street outside the hotel and together she and Svante drag the luggage to the electric car and open the boot. Svante hoists up the suitcase next to the microwave oven, induction hob and all the groceries Greta can conceivably need during the coming two weeks. Then Roxy jumps in and settles down, Greta feeds the destination into the satnav, and they drive out of the car park.

'The battery range is twenty kilometres short,' Greta says while they slowly and almost soundlessly roll towards the highway.

'Today we'll just drive on the battery,' Svante says. 'Slowly. We'll minimize electricity consumption and see how far we get.'

They drive on towards Kalix and turn north towards Gällivare.

The summer landscape zips past the car window at 80 kilometres an hour. The forest looks different to them now. A few years back they saw trees and nature and untouched fields.

Now they see clear-cutting, plantations and monocultures that have deprived the soil of its diversity and resistance.

Greta would have preferred to take the train, because no personal vehicle in the world can be considered sustainable, however electric it may be. But it's still an impossibility because of her eating disorders and compulsions, and the simple fact of being able to travel like this is such enormous progress that would have been unthinkable even a few months ago. Greta's energy has increased a little every day since last spring. Since the writing contest in the newspaper *Svenska Dagbladet*. Since she started planning her school strike.

Farms are sunk in 27°C summer heat. Abandoned farms. Working farms. Farms with livestock and people.

Farms with heaps of broken-down motor vehicles. Cars, tractors, campers, snowmobiles, snow-blowers, mopeds and motorcycles. Every other driveway is a potential motor museum.

But in some places, just beyond the road, you can still sense the dream of a different, simpler and perhaps better life. Small red houses rise out of the scraggy earth and give shape to the silhouette of an era that almost seems to have been frozen in time.

They are listening to *This Changes Everything* by Naomi Klein. Periodically they pause the audiobook to talk about what she's saying. Then they back up and listen again.

Play, listen, pause.

The bushes, thickets and the green pine forest extend almost all the way up to the Arctic Circle and an exotic white road sign announces that you are crossing the 'cultivation boundary'. The road is almost perfectly straight and deserted. Mile after monotonous mile. Scrawny trees that slowly shrink in size the further from the Gulf of Bothnia they get.

Some kind of black, long leaves are hanging from the pines;

they don't know what they are. Greta takes some pictures and says that they can ask someone tomorrow when they arrive.

Driving conditions are perfect for the electric car and the battery's range gets longer and longer relative to the destination.

'But we'll stop and charge in Kiruna anyway. We have to buy bread and vegetables, don't we?'

'Mmm,' Greta answers.

At the Co-op in Kiruna a sign boasts of 'Steps Forward for the Climate'. So far it consists of two electric-car chargers in one corner of the enormous car park at the Kiruna shopping centre. One of them is broken and plaintively blinking red, but the other one works and Svante politely asks the person idling there in an SUV if they can wait at one of the other vacant spots so he can get out of the car and plug it in. The car charges at 50 kilometres an hour and they walk slowly across the crowded car park towards a grove where Roxy gets to run off some energy and sniff the surroundings. She is far from County Cork and the backyard where she grew up.

It's hot in Kiruna too. Just as hot as in Luleå. It reeks of exhaust, frying oil and freshly cut grass. A man comes out of the DIY store with a hedge-trimmer under one arm. He is wearing cut-off jeans and a white T-shirt, and under his other arm he is carrying an empty box, which he sets down right outside the shop door. He is ready to go out into the neighbourhood and give the Arctic vegetation what it deserves.

They use the toilets at Burger King, squeezing through overflowing trays, chairs and tables loaded with Whoppers, Coca-Colas and French fries. The floor is sticky with ketchup and spilled soft drinks.

Men in hiking gear are standing outside the state off-licence with beer crates, fishing rods and backpacks spread out on the pavement in front of them. In anticipation of being in the wilderness they have transformed their little corner of the shopping centre into a men's changing room – they swear, spit and laugh loudly. It's fishing time. Time for the great outdoors. And time to drink.

Greta and Svante buy what needs to be bought and then drive on. The radio is playing Beata's favourite song, 'Whatever It Takes' by Imagine Dragons, and Svante misses her so much it hurts.

Wishes she could be there too.

Wishes they could do everything together.

Beyond the huge iron-ore mine there's a glimpse of the mountains and all that was once untouched. Svante tries to point out where the new, relocated Kiruna will be, but it's all very unclear. On the right, all that is visible is a big grassy slope and high-rises designed by the architect Ralph Erskine.

'I think the town is going to be somewhere over there behind the hill, but I'm not sure,' Svante says. 'A big part of the town will be moved, because it's no longer safe. The LKAB mine has grown so much and they've extracted such huge quantities of ore that everything is about to collapse. So now the state-owned mining company is boasting about how they're going to foot the bill for the whole move.'

'It's the least they can do,' says Greta.

'Yes, mining operations aren't exactly a non-profit enterprise,' Svante says, pressing start on the Naomi Klein again while they follow the Iron Ore Line to the northwest. After a few kilometres they brake for a herd of reindeer, and Greta takes pictures with her cracked old phone which, for a whole year, had served as a wifi router for the asylum-seeking family that was living in our house on Ingarö.

*

They are already in another world – a world where cars still have to adapt to animals. The reindeer cut a path around the oncoming trucks until finally our car can slowly move through the herd and continue across the plain towards Torneträsk.

Scene 88. A Time Machine

Svante is thinking so hard his whole body is vibrating. He would really like to be able to answer the question. He wants to show that he is well enough informed on the subject to justify his presence among the two dozen university interns from all over Europe who fill the classroom at the Swedish Polar Research Secretariat in Abisko. But the question is a difficult one.

'What is the efficiency rate of solar panels?'

No one can answer, even though the question concerns renewable energy and even though all those present are studying sustainability, ecology, biology or the climate. Svante considers taking a stab at it – it has to do with gradient and degrees, he thinks – when Keith Larson, the evolutionary ecologist currently at Umeå University, unexpectedly points towards Svante and Greta's desk.

Roxy is sleeping under Greta's chair and Svante feels a little surge of stress before, out of the corner of his eye, he sees Greta raise her hand. He has no time to react.

'Sixteen per cent,' she answers loudly and clearly in English, the first time in several years that Svante has heard her speak of her own volition with someone outside the family besides her teacher Anita.

He doesn't understand where it came from. Much less does he understand the origin of the answer, the 16 per cent.

'Exactly!' Keith Larson responds happily from up by the whiteboard, repeating the answer one more time. 'Sixteen per cent!'

The students from all over Europe look with amusement and surprise at Greta while the lecture continues up by the screen.

Afterwards they stand on the roof of the measurement station and Keith Larson explains that the major changes started to pick up speed in the late eighties. Since then they have been moving fast. Extremely fast.

'The snow, ice and glaciers had a delaying effect up here in the Arctic. But once they'd melted a certain amount it has gone even faster.'

Keith is an American, but nowadays lives year-round at the compound, which is the world's oldest, still-functioning research station. It was built at the same time as the Iron Ore Line, and now a unique project is being carried out in which research from a century ago is being repeated. Along the slope of Nuolja mountain the vegetation is being measured at exactly the same place now as then.

And even if the research results are still far from ready, major differences can already be seen. Some with the naked eye.

'Of course what is happening here follows the trends from across the whole planet. The temperature is rising and the difference is greatest close to the poles. The tree-line is climbing up the mountain slopes, the vegetation zone follows and the Alpine environment is shrinking. When the temperature increases, trees and bushes can grow higher up, where before it was too cold.'

Another kilometre-long ore train thunders through the landscape between the research station and the mountain tops.

'Here in Abisko the results of climate change manifest differently than in most other places. Here, everything is obvious even for those who don't have any experience studying these types of changes. Just look at the transition zone, where the bushes grow, which today is four times the size it was a hundred years ago.'

Keith Larson explains that perhaps the biggest problem is that the Alpine zone is shrinking. This means the species that live there are forced out by other animals, insects and plants that are following the trees' progress up the mountainsides. They have to keep moving until there is nowhere else to go. The balance is upset. The conditions are changed.

'If you look at the tree-line on Nuolja you see that it has climbed up the mountain slope, just like it has at countless other places around the world. Now, a ski lift has been built there too, which means that the reindeer haven't grazed that area, but the result is much the same,' he says, pointing up towards the mountain. It is hot up on the black roof and they decide to move down into the shade to talk. But before they go down and leave the view up towards the mountains, he points out the most striking change.

'Fifty years ago the tree-line was at the same place as it was when Thore Friis did his research studies a hundred years ago. But now it is moving. Faster and faster. Today it is 230 metres further up the mountain slope.'

'Did you say 230 metres?' Svante repeats.

'Yes,' Keith replies. 'This is the front line for the Arctic. The changes here are happening at high-speed, as I said. And I'm surprised that more researchers don't make their way to Sweden and Abisko. This is such a unique environment and an incredible amount is happening.'

The next morning Greta, Svante and Roxy accompany four German university students who are collecting data from the same area of the mountain that Friis used for his study between

1916 and 1919. Yrsa, the daughter of our book's publisher, goes with them too. She has a summer job as a communications officer for Keith Larson and his team.

'It's unique that such detailed studies are still preserved today and can be used in our research a hundred years later. It's like a time machine,' the students explain.

They start close to the top by setting up a tripod along the path and registering their position with their iPads.

Greta and Svante follow a short distance behind. The view is entrancing and you can see for an incredible distance: Torneträsk. The mountains. And the Iron Ore Line, of course, which never rests. Far below they see the trains churning on towards Narvik and Norway. Towards the harbour, the ships, the sea and all the waiting industries around the world.

High up on the mountain it's late winter. Further down comes spring, with flowers and rippling streams. At the transition zone it's summer and it's time for lunch. The flies buzz and there is a scent of flowers and moss. It is calm and everyone takes off their jackets and fleeces.

'I assume that this is what it's like for you every day,' Svante jokes.

'Not really,' they say with a laugh in return.

Greta sits down a few metres from the others. She takes out her glass container of boiled bean pasta and her fork, and takes a deep but almost imperceptible breath. Then she starts eating.

It is the first time she has eaten in the company of strangers in almost four years.

She is making her way towards the time before OCD and eating disorders.

Or rather.

To the time after.

Scene 89.
Tropical Nights

'I've lived here for over thirty years,' the woman from reception says as she sets out fresh porridge for the breakfast buffet, 'and I've never experienced anything like it. The temperature was over twenty degrees the whole night. That must be unique.'

'If the temperature is over twenty at night, I think that's called a tropical night,' Svante replies. 'That must not happen very often north of the Arctic Circle,' he says with a laugh, so as not to repeat the mistake made at the last hotel.

But here the heat is not received with the same unchecked enthusiasm as down in Luleå. Here there is cautious concern about the extreme summer temperature, and the staff don't really know how to answer the guests' questions about where one can hike in the shade, how hot it will be on the bare mountain top today, and whether it's possible to make it up to Lapporten in this heat.

Svante fills a bowl with porridge and refrains from asking for oat milk, because the odds that there would be any don't seem strong enough for him to risk being seen as someone who thinks they're a little better and more sophisticated than everyone else.

Someone from Stockholm, in other words.

As if that wasn't already obvious from the extension cord between the makeshift charging point and the electric car that is itself squeezed between the parking lot and the restaurant's little wooden deck.

Greta eats in the minimalist bedroom with Roxy. Pågen's

wholegrain rye, sourdough and lingonberry bread, as always. Plain.

It's hot on the restaurant patio. The butter melts on the bread and Svante might as well have been sitting in Rome or Barcelona. He pours his fourth cup of coffee as the last guests depart, and the hotel staff sit down at the table next to him and take their morning break in the sun.

They talk about the heat, not unsurprisingly. And who said this and did that in the village. There are four women, one of whom is from further south. From Hälsingland. She seems to have the hardest time handling the heat and they poke fun at her. She ought to be used to it, and so on. She's the same person who advises the guests about where they can find shade when they're out hiking in the heat.

'Almost best to stay indoors,' she says, without the slightest hint of irony.

The roar from a helicopter taking off or landing on the helipad above drowns out the conversation from time to time. The air smells of kerosene, fresh coffee and pressure-treated wood.

After a while a man joins the women. They are well acquainted. He's a helicopter pilot and the conversation drifts to what it's like at the other hotels and cabins around the mountain area.

'It was twenty-five degrees at Kebnekaise yesterday,' and 'Have you heard that the south peak is no longer the highest, because the glacier has melted so much?'

It wasn't news.

Business is good in the helicopter world, but he's far from content. He wonders whether he could increase the profitability for everyone if he lowered his prices sharply so that in theory he'd be losing money on fuel and the helicopter, but if everyone pitched in to cover the costs it would mean they could get many more people out to the cabins, and more people in circulation means more income for everyone.

They all think it's a great idea.

Svante is of a different mind, but he keeps his thoughts to himself.

This is a part of Sweden that has paid for almost everything. The ground has been dug out. The rivers have been tamed and the forest cleared. And the money has always ended up in wallets further south. Fat wallets.

Very fat.

Greta and Svante prepare their lunch and finish packing for the day's hike. They walk across the hotel's gravel yard. Roxy runs ahead and pauses for eye contact every ten metres. The staff's breakfast is over and three of them are gathered around the air-source heat pump outside the restaurant, passing the user instructions from hand to hand. They read out loud to each other and cautiously finger the panel with its buttons and digital display.

'It should be able to pump cold air too,' the woman from reception says. 'Like air conditioning.'

The thermometer outside the supermarket shows 31.7°C and they don't really know where to go. They're panicking slightly. Up on the bare mountain top it's probably cooler but there's no shade. The snow on the peaks all around has all but disappeared in three short days.

It ought to be cooler along the river, they venture, and they're right. There are also patches of shade to be found in the low-growing mountain forest. Periodically they stop by the rapids and splash themselves. The trees, the ground, the grass, the plants and the marsh smell like nothing they've ever smelled before. After several days in the intense heat a new environment has emerged. A brand new world with new scents, new colours and new living conditions. Sometimes they stop and get on their knees, noses to the earth and the moss, and just smell.

They take a break by the river beneath some white boulders. The water is green and white from the rapids, which are strong along the middle channel and against the opposite shore. If Roxy were swept away by the current she wouldn't stand a chance. She'd be taken down the river and the falls five kilometres to Torneträsk. So they stick to a little cove near the shore.

The river is cold. But not too cold. They take a dip and drink the mountain water they're swimming in. They guzzle it down. They dry themselves in the sun on the stones until it gets too hot. Then they jump in again.

Greta finds a black stone on the shore that's shaped like a heart. A perfect, coal-black heart.

'Like Knight Kato,' she says. 'A heart of stone. We ought to throw it into the water like in *Mio, My Son*.'

'Go ahead,' says Svante.

But Greta hesitates.

'But imagine, it took millions of years for this one stone to end up here on the shore. And what if it would make some other person passing by happy to see it?'

'Eh,' Svante says quickly, 'we humans have so much to be happy about and grateful for. We don't deserve more.'

Greta clutches the black stone and throws it with all her might out into the middle of the river. Roxy perks up and is about to chase after it, but stops on a dime and stays on the shore and watches the ripples from the splash disappear into the rapids' whirling roar.

Scene 90.
Something Very Big and Unexpected Has to Happen

The next morning it's much cooler. It's drizzling outside and a different weather system has moved in over the mountains. They take the trail up towards Trollsjön Lake and Roxy zig-zags up and down the mountain slopes. The landscape is a cross between *The Sound of Music* and *Lord of the Rings*, with gigantic stone blocks resting in the green grass between the rock walls that reach into the sky like skyscrapers on both sides of the valley. Yellow flowers are blooming everywhere.

Greta's energy increases with each passing day. She talks about the school strike and keeps asking how to go about it.

'Whatever happens, you have to do it all on your own,' Svante says for at least the tenth time. 'You have to be able to answer every question. And you have to know all the arguments and answers. The journalists are going to ask you about everything.'

'What are they going to ask?'

'The same things I said before,' Svante replies.

'But tell me more. What can they ask? Question me like you're one of them.'

' "Did your parents put you up to this?" You're going to get that question all the time.'

'Then I'll tell it like it is. I'm the one who influenced you and not the other way around.'

'Exactly,' Svante replies.

Greta continues, 'They can just look at my Twitter account and see what I've written. I might be shy and unsociable, but it's not like I've been living in a vacuum. I've won second prize in a writing contest. I've got publishers to rewrite textbooks. You can read about that online.'

'But they're not going to. Unfortunately. Only the haters dig up the past. No one else cares. And if it doesn't fit into their story I can guarantee they aren't going to bring it up. But people will understand. Your climate struggle is no secret. There's even a whole pitch for a TV programme about how you got your mother to become an "involuntary environmental warrior", and because it came from a producer and a production company that in theory can do whatever they want, I guarantee that every decision-maker at Sveriges Television has read it.'

Greta takes in what her father is saying.

'Nothing came of that programme, right?'

'No, we can let that one go. It's been a year and a half. The public broadcasting service won't touch the climate issue with a ten-foot pole.'

'But what else are they going to ask?' Greta continues.

'All kinds of things. The important thing is that you tell it like it is and emphasize facts. You have to have all the facts and make sure that you know what you're talking about. They're probably going to ask you, "But what are we supposed to do about it?" or "What is the priority?", because we grown-ups have learned that we must always have concrete answers to every question, even if we don't. How something is said is more important than what is said. Keep that in mind.'

'Okay,' Greta says slowly. 'But there are no solutions within the current system. The only thing we can do is start treating the crisis like a crisis.'

'Exactly,' says Svante. 'But no one is going to understand that. So you'll have to repeat it. Again and again and again.'

Just like me, Svante would prefer that Greta drop the idea of a school strike. It would definitely be easier that way. But he also sees how much energy it gives her to talk and think about it and he tries to answer all of her questions. However difficult they may be.

They leave the path to walk in among enormous stone blocks. They climb up to a perfect lunch spot under a boulder that acts as a shelter from the sporadic rain showers.

Svante sends a picture to Beata and me on his phone. It is still an incredibly big deal for us that Greta can eat in new places. Outdoors even. Bean pasta with a pinch of salt and slices of lingonberry sourdough rye can be taken almost anywhere, and that opens up amazing possibilities.

Like hiking in the mountains.

The cloud cover breaks up and the sun peeks out again. Opposite them, the entire mountain wall is like a gigantic, improvised waterfall. Water rushes for several kilometres down the cliff face.

They look out over the valley another few hundred metres below. A small river delta spreads out across the luxuriant grass and hundreds of reindeer are moving about like ants down there. Perhaps there are thousands.

Suddenly some at the edge of the herd start running and a few follow. After a while they slow down, stop and continue grazing.

'Going on school strike for the climate is going to be incomprehensible to anyone who doesn't understand how serious the situation is,' Greta says happily. Almost euphorically. 'And because almost no one knows, almost no one is going to understand. I am going to be so incredibly hated,' she says with a laugh.

'Maybe kids will understand,' Svante says.

'No. Kids act like their parents,' she answers. 'I haven't met a single kid who cares about the climate. Everyone says that it's the kids who will save us, but I don't believe that.'

Svante sits silently. He hopes she's wrong.

Greta continues, 'If we have two years left before the emissions curve has to go down, something has to start happening now, and by next spring something must have happened. Something huge and totally unexpected.'

The reindeer move slowly around the river down in the valley. The air is warmer now.

They pack up and continue the hike until they are at the foot of the last hill before Trollsjön. Once they reach the lake, they should be able to see the bottom thirty or forty metres down because the water is so clear. They see how the mountain walls around the lake are dripping from rain and melt water and everything is in constant motion. The wind is gusting.

You can sense the lake beyond the ledge above the path. But Greta looks tired.

'Can you manage?' Svante asks. 'There's only a few hundred metres left, we're almost there.'

'Don't know,' Greta replies.

They stand there. Take a picture with the phone. Wait.

'When I was little I was taught never to give up. You should always try harder,' Svante says, preparing a little speech.

'My first summer job was at a laundry in Bromma, and it took like an hour and a half to get there every morning. I laundered soiled sheets and blankets from a nursing home and I wanted to quit right away, but Grandma made me continue and I've always thought that it was a really good thing. Me not giving up. But. Now I'm not so sure any more. Sometimes I think that maybe we ought to give up a little more often. Or at least take a few steps back sometimes.'

It is starting to drizzle again and it is four kilometres down to the road. They've been gone for ten days and soon it will be time to drive back to Stockholm. Tomorrow they plan to drive to Kvikkjokk as the first leg of the journey homeward.

'You know what,' he says, 'let's turn around here. We don't need to see everything. We don't need to have been everywhere.'

Scene 91.
Every Dinosaur Had ADHD

I am so tired of our story.

But now we're sitting there, telling it all again.

Svante talks. I talk. We converse politely because the children are in the room with us.

Greta is investigating a cube and some educational triangles that are on the table in the consultation room.

Beata is squirming and rolling her eyes.

She wants to go home and dance. She's as tired of BUP as I am.

When we're done and the children have gone out ahead, the doctor sighs and shakes his head.

'Yes, good Lord,' he says, 'you do need help.'

All three of us smile. Everyone wishes us well.

Everyone does their best. And often a little more.

Exactly like most other people; people who want to do good things from their unique perspectives.

The whole family walks home together along Fleminggatan. It's summer. The birds are singing in the trees and summer clouds stretch across the sky like an upside-down archipelago.

An aeroplane has drawn a line over the horizon.

It's superfluous to us now. We don't need it any more.

Svante has promised to take Greta to a building supplier's to buy a scrap piece of wood that she can paint white and make a sign out of. 'School Strike for the Climate', it will say – she decided that a long time ago. And although Svante and I realize what enormous risks she is going to be exposed to – although more than anything we want her to drop the whole idea of going on strike from school – we support her. With just the right measure of enthusiasm. Because even though the school strike is drawing nearer and nearer she doesn't seem to be showing any signs of dropping her idea. Quite the opposite. And we see that she feels good as she draws up her plans – better than she has felt in many years. Better than ever before, in fact.

In one of the shops in the Västermalm centre looms a big green cloth dinosaur. As we hurry past we glimpse our reflections in the window: Beata, Greta, Svante, the dinosaur and me.

If we hadn't had so many diagnoses, so many obsessive-compulsive and eating disorders, if Svante didn't need to pee like he always does, we could have stopped and taken a photo.

It would have been a perfect transition to a larger geologic perspective.

But it is what it is.

'I wonder if the dinosaurs had ADHD,' Svante says.

'Yep,' Beata answers. 'They had Asperger's, OCD, ODD and ADHD like me. That was why they became extinct. They had too many thoughts in their heads and couldn't concentrate and so they went crazy from all the disturbing sounds.'

Scene 92.
Infinite Growth on a Finite Planet

The dinosaurs lived on earth for about 200 million years, a short period if you consider the earth's 4.6 billion-year history.

We humans have only existed for about 200,000 years. But we've already managed to create stuffed animals shaped like reptiles that died out over 60 million years ago; stuffed animals that we mass produce in China and ship all over the world and sell to anyone who can afford such things.

Not everyone can.

But many can, and every day they become more and more, and that sort of thing takes resources.

But the resources aren't increasing.

There are limits to what we can get out of this well-exploited planet every year.

One of those resources is running out at a furious speed, and the dinosaur in the toy shop shares part of the blame.

We all share part of the blame.

But not equally, of course.

The world's richest 10 per cent account for half of all greenhouse-gas emissions, which right now are destroying one of our most important natural resources: a balanced and functioning atmosphere.

At the current pace of emissions, that natural resource will soon be gone, and the fact that so few of us are aware of it must be one of *Homo sapiens'* greatest failures.

But how are we supposed to know?

We're in a crisis that has never been treated as a crisis.

The poorest half of the earth's population accounts for only 10 per cent of the world's CO_2 emissions, and if we are going to find role models, that is probably where we'll find them. Rather than among celebrities like me. Or among Hollywood stars and former American politicians with more annual flight hours than an average fighter pilot.

The climate researcher Kevin Anderson makes the point that if the world's richest 10 per cent were required to lower their emission levels to the European Union average, the world's emissions would go down by *30 per cent.* That – and many other quick measures – could give us time.

Scene 93. The Big Stage

We really thought she'd be home by lunch. If she even left at all.

But that's not what happened.

On the morning of 20 August 2018, Greta gets up an hour earlier than on a regular school day.

She has her breakfast. Fills a backpack with schoolbooks, a lunchbox, utensils, a water bottle, a cushion and an extra sweater.

She has printed out 100 flyers with facts and source references about the climate and sustainability crisis. The text is 5,303 characters long, including spaces. And on the front it says in big, black letters:

We children don't usually do
as you say.
We do as you do.
And because you grown-ups
don't give a damn about my future,
neither do I.

My name is Greta and I'm in 9th grade.
And I am going on strike from school for the climate
until Election Day.

She walks her white bicycle out of the garage. It has barely been used. In the past four years she hasn't had the inclination or energy to go places on her own. Much less to ride her bike for fun.

She gets on the saddle, glances behind her, and rolls off

along Kungsholms Strand, past the City Hall, and then, towards Drottninggatan.

On Tegelbacken, some tourists are standing around smoking as the old steamboats send up their coal-black clouds into the clear blue late summer sky above the rush-hour traffic on Centralbron and Söderleden. Svante cycles a few metres behind her, with the sign under his right arm.

The Thursday before – four days ago – she walked past to see what the streets around Parliament look like and to decide where she would sit.

'Against the wall inside the pillars would be good,' she said.

Svante nodded.

Then she asked him to take a picture of her by the railings in front of the bridge.

She had on a black T-shirt with a crossed-out aeroplane on the front.

Like a road sign.

Before they left she stopped for a moment by the statue of the begging fox with the blanket.

She looked at Drottninggatan. At the bridge. She observed the Parliament building on the other side of Stockholms Ström.

'Is it really true that no one else has ever done this before?' she asked.

'No, I don't think so,' Svante answered.

'But it's so simple,' she said.

Then she cycled home and finished painting the white hardboard sign. The sign she bought from the leftover pieces at Bygg-Ole for 20 kronor.

The weather is rather lovely this Monday morning. The sun is rising behind the Old Town and there is little to no chance of rain. The cycle paths and pavements are filled with people on their way to work.

As cycle paths and pavements are on an ordinary weekday morning at the end of August.

Today is the start of school.

Outside Rosenbad, which houses the prime minister's office, she stops and gets off her bicycle.

Svante helps her take a picture before they lock the bicycles to the railings and hang Greta's bicycle helmet on the handlebars. Then she nods an almost invisible goodbye to Dad and, with the clunky sign in her arms, staggers around the corner where the bicycle path turns left towards the government block.

'Hurry off to school now, okay?' Svante calls. Half joking.

She doesn't react. Just keeps going.

And there, somewhere on the road towards the bridge by Drottninggatan, she passes the invisible boundary, the point of no return.

She crosses the bridge and passes under the arch, continues a few steps onto Riksgatan before she stops and leans the sign against the greyish-red granite wall.

Sets out her flyers.

Settles down.

She asks a passerby to take another picture with her phone and posts both pictures on social media before she puts the phone away in the small violet Björn Borg backpack, a Christmas gift from her grandmother four years ago.

Svante stays by the bicycles until she has disappeared from view. A large salmon jumps up out of the water and for a moment is suspended in midair, before it breaks the surface with a splash.

A few hundred metres above Helgeandsholmen a bird of prey circles, round and round.

Maybe an eagle.

Or an osprey?

He lets go of the railings and walks over to Fredsgatan, up to a coffee shop by the Ministry of Education. Orders a large oat-milk latte, sits down by the window and tries to work.

But it's not an easy thing to do.

After a few minutes the first sharing on Twitter starts. The musician and climate advocate Staffan Lindberg retweets her post.

Then come another two retweets.

And a few more.

The meteorologist Pär Holmgren. The singer-songwriter Stefan Sundström.

After that, it accelerates. She has fewer than twenty followers on Instagram and not many more on Twitter.

But that's already changing.

Now there is no way back.

A documentary film crew shows up.

It's the filmmaker Peter Modestij, who found out what Greta was planning the week before, by chance, when he called me to discuss a film script he was in the process of writing. The whole family had read his script last winter and he was eager to have our input because he could see that Greta was very much like the main character in his upcoming feature film.

Now he has convinced a film company, on spec, to pay for two days' filming of Greta's school strike.

Peter's friend Nathan Grossman, who made a documentary on pigs and the meat industry for SVT with the comedian and actor Henrik Schyffert, has come along. He says hello to Greta and asks if she's okay with them filming.

She has nothing against it and they put a microphone on her.

The cameras roll. As of now, everything that is said and done will be documented in sound and image.

But Greta couldn't care less about their presence. All she's thinking about is sitting there and seeing what happens.

So she sits there.

Alone, leaning against the big wall.

No one stops.

One or two people cast a worried glance at her, but the majority choose to look the other way.

They have more important things to do.

And isn't this all a bit awkward?

Two middle-aged women stop and explain that school attendance is mandatory and that she should focus on school. They express their worry for her future and for her continued studies.

A middle-aged man named Ingmar Rentzhog stops by and introduces himself. He films Greta and asks if he may post a video and a picture on Facebook.

She nods.

Meanwhile, Greta's Twitter and Instagram posts have started to go viral.

Svante calls and tells her that the newspaper *Dagens ETC* has been in touch with him and they are on their way. Right after that *Aftonbladet* shows up and Greta is surprised that everything is moving so fast. Happy and surprised.

She wasn't expecting this.

The photographer Anders Hellberg from the environmental magazine *Effekt* arrives and starts photographing. He walks around searching for angles. But mostly he stands in the middle of the street where people are passing by.

He stands there with the camera in one hand and smiles. Hour after hour.

'This,' he says, when a few more people have stopped and started talking around Greta. His face and his camera are turned towards the scene playing out in front of him.

'This!' he repeats over and over again. Then he lets out the happiest of laughs.

Like him, quite a few are stopping by. People who for decades have toiled and struggled to draw attention to the climate crisis.

Ivan and Fanny from Greenpeace show up and ask Greta if everything is okay.

'Can we help with anything?' they ask. 'Do you have a police permit?' Ivan asks.

She doesn't. She is 'on strike from school' and didn't think a permit would be needed.

But evidently it is.

'I can help you,' Ivan says, and gives her a short explanation about the rights and freedoms of democracy.

But Greenpeace is far from alone in offering their support.

Everyone joins in.

Everyone wants to do their utmost to help out.

But Greta doesn't need any help.

She manages all by herself. She is interviewed by one newspaper after the next.

The simple fact that she is talking to strangers without feeling unwell is an unexpected joy for us parents.

Everything else is a bonus.

Svante gets a link on his phone to the first interview and reads what it says in *Dagens ETC*.

He reads it again.

And he doesn't understand how it has happened but it's the best interview on the climate that he has ever read.

Greta's answers are crystal clear and cut right through the noise.

As if his daughter had never done anything but being interviewed by journalists.

On the other hand she hasn't eaten lunch that day.

She hasn't had time. And it was hard with everybody watching.

This is a big problem, but she did eat once she got home in the afternoon.

Before it's time for her to jump on her bicycle, a journalist from Radio P3 news introduces himself and tells Greta that her posts have been shared like crazy today.

'Okay,' says Greta.

'I mean, like *crazy*,' the journalist clarifies. 'Is it okay if we ask a few questions?'

It's way past three o'clock and the school day is long over.

'I'm sorry,' Svante says, 'but I think she's pretty tired.'

'It's okay,' Greta interrupts.

And so she does one more interview before she cycles home.

She's happy. Her whole body shows it. It's as if she is bouncing on the bicycle seat as she rolls away.

Scene 94.
A Movement

It's been said that the moment a person acting alone is joined by someone else, a movement begins.

In that case the global school-strike movement was founded at approximately nine o'clock on Day 2 of Greta's school strike. At least that is when Mayson, an eighth grader at Adolf Fredrik's Music School, asks whether it's okay if he sits down and keeps Greta company. Greta nods.

And from there on out she never sits alone again.

Another two schoolgirls walk up and sit down on the cold cobblestones.

A student from Stockholm University.

A French teacher in his thirties who didn't go to work that day and instead came all the way to Stockholm from Gothenburg.

'I'll probably get fired,' he says. 'But it doesn't matter, because something has to happen. Someone has to do something.'

Then *Dagens Nyheter* and TV4 are there. Greta's teacher joins her and is interviewed on the news broadcast.

'As a teacher I can't support this,' she says. 'But as a fellow human being I understand why she's doing it.'

The segment is edited in a way that makes her out as supportive, and during the coming weeks she is bullied and shunned so much at her workplace that she is forced to go on sick leave.

The first haters start to attack, and Greta is openly mocked on social media. She is mocked by anonymous troll accounts, by right-wing extremists. And she is mocked by members of

Parliament from parties that many of her closest relatives vote for. Members of Parliament from parties that an overwhelming majority of her neighbours vote for.

It's apparent in the eyes of people we meet on the street, or while we're doing the shopping.

The politicians' studied, ridiculously derisive comments are seeds that are carefully planted in fertile ground on social media and quickly grow into sturdy stalks of the deepest hatred and contempt. But that's no surprise.

On the other hand, Greta did not count on the hate and derision coming from individuals close to our own family. From close family members even.

'If you don't have full insight into the climate crisis, what I'm doing will obviously seem totally incomprehensible and I know that no one actually has any idea about the climate crisis,' she says over and over again.

She says many things over and over again.

Like a mantra.

'The school strike is non-partisan and everyone is welcome,' she repeats for the umpteenth time to a passerby who asks if it's a political thing.

Svante stops by to make sure that everything is okay.

He does this a couple of times every day.

Greta stands by the wall and there are a dozen people around her. She looks stressed. The journalist from *Dagens Nyheter* asks whether it's okay if they film an interview, and Svante sees out of the corner of his eye that something is wrong.

'Wait, let me check,' he says, and takes Greta behind a pillar under the arch. Her whole body is tense. She is breathing heavily, and Svante says that there's nothing to worry about.

'Let's go home now,' he says. 'Okay?'

Greta shakes her head. She's crying.

'You don't need to do any of this. You've already done more than anyone else. Let's forget about this and get out of here.'

But Greta doesn't want to go home. She stands perfectly still for a few seconds. Breathes. Then she walks around in a little circle and somehow pushes away all that panic and fear that she has been carrying inside her for as long as she can remember.

After that she stops, and stares straight ahead.

Her breathing is still agitated and tears are running down her cheeks.

'No,' she says. At the same time she lets out a neighing sound. Like a call. An animal. Everything is in the balance.

Wobbling.

'No,' she repeats.

'Do you want to stay?' Svante asks carefully. 'Are you really sure?'

Greta wipes away the tears and grimaces.

'I'm doing this,' she says.

She turns around. Her body is calm. She gives the journalists who are waiting on the other side of the pedestrian street a relaxed smile.

Greta goes back to her strike and Svante follows her every movement. He stands behind the pillar for over half an hour, observing his daughter. He thinks that at any moment she is going to run away. At any moment she'll be overcome by stress and fear.

But nothing happens.

She simply stands there, speaking calmly with the journalists. One after another.

Svante thinks that she surely must be feeling terrible, and that she ought to turn round and get out of there. But Greta doesn't turn round.

She stays in the middle of the crowd.

Now and then she lets her gaze wander along the facade of Parliament. She looks calmer now than during the first day, and anyone looking closely can see that she's smiling, an almost invisible smile. As if she knows something the rest of us don't.

Then when the journalists have left she settles down on her little blue cushion and reads her books so as not to fall behind in her schoolwork.

She's reading *My Mother Gets Married* by Moa Martinson for Swedish literature. In her social studies textbook she's reading about how parliamentary elections are conducted and how government, Parliament, committees and ministries work.

She's reading in the biology book about genes and hereditary characteristics.

She only uses the phone to post the day's strike image on Twitter and Instagram, because she's on school time *and during school time you mustn't use your mobile phone.*

At three o'clock she packs up and cycles home.

Scene 95. The Third Day

We monitor how Greta is feeling as closely as we can. But no matter how we look at it, we can't see any signs that she's feeling anything but good. Better than good, even. She sets the alarm clock for 6.15 a.m. and she's happy when she gets out of bed. She's happy as she cycles off to Parliament, and she's happy when she comes home in the afternoon.

During the afternoons she catches up on schoolwork and checks social media.

She goes to bed on time, falls asleep right away, and sleeps peacefully the whole night long.

Eating, on the other hand, is not going well. Not during the strike, anyway.

'There are too many people and I don't have time. Everyone wants to talk all the time.'

She takes bean pasta with her but it's hard to make that work.

She has to have extra snacks when she comes home in the afternoon.

'You have to eat,' Svante says. 'It won't work if you can't eat.'

Greta doesn't say anything.

Food is a sensitive topic. The most difficult one. It's been that way for several years and there's no real solution in sight. But on the third day something else happens.

Ivan from Greenpeace stops by again. He's holding a white plastic bag.

'Are you hungry, Greta? It's noodles. Thai,' he says. 'Vegan. Would you like some?'

He holds out the bag and Greta leans forward and reaches for the food container.

She opens the lid and smells it a few times.

Scans the food with her nose.

Then she takes a little bite. And another. No one reacts to what's happening. Why would they? Why would it be remarkable for a child to be sitting on the ground, with a bunch of other people, eating vegan phad thai?

Greta keeps eating. Not just a few bites but almost the whole serving, and the scene that plays out there on the cobblestones in front of the school-strike sign changes everything. The usual manual goes into the bin and the map is redrawn.

A little later a man comes fully loaded with bags from a big hamburger chain. He doles out hamburgers, French fries, ice cream and soda to anyone who wants it.

'These are vegan and vegetarian hamburgers,' he says proudly, setting half a dozen paper bags with the company's logo in among the children.

'I don't think this is a good idea,' Greta says, and tries to explain to the children. But Greta speaks too softly, the children are too hungry, and her message does not get through.

Everything is eaten up.

When Svante walks past to check that everything is okay, the food has been consumed and the man is talking happily with everyone who has gathered. Svante introduces himself, takes him aside and explains.

'Greta has explicitly said that she absolutely does not want any sponsorship, so I would like you to remove the bags and not offer the children food when they are on strike.'

'But what will they eat?' he asks.

'I'm sure we'll work it out,' Svante says. 'But because there are a lot of cameras here, Greta doesn't want anyone coming along and setting out their products, because that would feel wrong. She has talked about this before.'

Svante explains the guidelines Greta has developed: No sponsorship, no advertising and no political-party logos. The man gets a bit cross, and starts talking about how much his company has invested in vegetarian alternatives and that their hamburgers are climate-neutral because they invest a great deal in planting trees in East Africa. He says that he has worked with sustainability issues for over twenty years.

'Yes, but it's still the case that you're here during working hours representing a company whose primary revenue source is and always has been slaughtering cattle and selling the dead animals' meat in a fast-growing hamburger chain. And this has nothing to do with the children who are on strike for the climate.'

'Sure,' he says, 'but people have to eat and we're all part of the same system.'

He points at Svante's shoes.

'You're wearing trainers. That's not sustainable either,' he says.

'No, but you can't really compare the fact that I have a pair of running shoes on with being sponsored by an expanding burger chain that earns hundreds of millions selling fast food.'

The man gathers up his bags and cups, and leaves.

After the scene with the hamburgers, Greta forbids Svante from coming anywhere near the strike. She wants to be on her own and doesn't want anyone speaking for her.

Greta flips to the chapter about the Swedish constitution in the social studies textbook and settles down next to the black-and-white school-strike sign.

Some soldiers from the Royal Guards stroll past. Young

men and women in camouflage uniforms, each with a little Swedish flag sewn high up on their jacket sleeves. They see Greta but look pointedly in the other direction. As if to underscore that in their world there is still no doubt whatsoever about who defends whom.

In the afternoon the man with the hamburgers asks Greta on Instagram if it's really true that she doesn't want him to offer food to the children who are striking with her.

'You're welcome to offer us food,' she responds, 'but it must be food that doesn't come from a company that you work for.'

In that case, he replies, it'll probably be hard for him to find the time.

Scene 96. Stronger and Stronger

I promise that any parent whose child hasn't talked to people for several years and who could eat only a few things in a few predetermined places will be extremely happy to see those complications vanish. I promise that, as a parent, you'd perceive the change as extremely positive. Almost like a fairy tale. Like magic.

Regardless of what any older conservative white men and women write on social media or in their newspaper columns.

Some think that 'someone' is 'behind all this'. A PR agency.

But that's not the case.

Greta's summer did not pass by in a series of clandestine meetings behind thick curtains at murky advertising agencies, where she was drilled in falsifying her background, her values and opinions. All under the influence of globalists, cunning left-wing economists and George Soros. That sort of thing.

All to reinforce government influence and increase our joint tax burden; all in the name of creating the eco-fascist, global super-state.

Each conspiracy theory is worse than the next.

Greta has not sacrificed four or five hellish years to simulating various life-threatening difficulties in order to now launch the world's most cunning PR coup.

On the other hand there are a countless number of people who stand behind her. Everyone who has struggled and toiled for decades to bring attention to the climate crisis is joining in. Like they've always done.

The channels are there from day one. And for some reason they seem to work better for her than the great majority of others who have got the same kind of attention before.

Everyone stands behind Greta.

Just like Greta stands behind them.

Everyone supports everyone.

'The reason that this is getting so big is probably because this is the most important question humankind has ever faced and because it has been totally ignored for over thirty years,' Greta says.

But none of the doubters are listening to anything she's actually saying. They're completely indifferent to the sustainability issue.

Scene 97.
In the Spotlight

Greta's energy isn't growing by the day.

It's exploding.

There doesn't seem to be any outer limit, and even if we try to hold her back she just keeps going. By herself.

After a full day of interviews in front of Parliament, she insists on taking part in a panel discussion at Kulturhuset, the Stockholm House of Culture. She bikes home and eats, then cycles back to Sergels Torg and practically runs up the escalators to the seminar. The place is packed. Greta is fitted with a microphone and steps on to the stage. She is received like a rock star and stands poised in the spotlight next to the politician and meteorologist Pär Holmgren, Professor Emeritus Staffan Laestadius and the policy spokespeople for Sweden's two biggest political parties.

Greta is given the floor and she tells it like it is, never mincing words.

'We find ourselves in an acute crisis and nothing is being done to handle that crisis.'

Staffan Laestadius says the same thing.

These are rock-hard words from the stage.

Unconditional words.

The atmosphere becomes both hopeful and ominous.

This is a new story being told – although the content and words are the same as before.

'It is absolutely this serious,' Pär Holmgren adds. 'I've been saying so for over ten years, and to be honest I no longer know if we're going to be able to solve this crisis. But, as I always say, it's never too late to do as much as we can.'

The male politician instinctively reacts with anger. He is furious and feels provoked by what has been said so far.

'We must focus on infusing people with hope,' he says, distancing himself firmly from what he has heard this evening.

The female politician reacts differently. She starts crying, sobs helplessly behind her cupped hands. Finds no words.

None of this is expected.

She takes out a handkerchief and for a brief moment she is at a loss. In the audience, Svante thinks that there has finally been a genuine, fully human reaction.

The pattern is being broken and that's hopeful somehow.

He wants to see her stay in that moment.

He wants to see what would happen if she gave in and perhaps dared to stare down into the abyss without flinching.

He wants to see what would happen if she gave herself that time.

If we all dared to admit our respective failures.

Gave everything a chance to stop.

But she pulls herself together. She sets aside the handkerchief and starts talking about our common challenges; about opportunity, jobs and green growth.

Green, eternal growth.

On the down escalators, Greta turns to Pär Holmgren and says, 'My God, it's worse than I thought. They really have no idea. The politicians don't actually know anything.'

'No,' Pär says and thinks for a few seconds. 'I think they are so used to dealing with representatives from the business world and lobbyists. The ones who always have answers to every question. The ones who always say that everything can be solved.'

'It's as if the politicians always have to be able to answer questions and can never say that they don't know. Even though they don't have a clue.'

'That's probably right,' Pär says in his quiet way, with a laugh.

'But that's really insane,' she says.

And it is.

Svante and Greta walk their bikes past the Åhléns department store and over towards the Klarastrand viaduct.

'It's like everyone is obsessed with hope. Like spoiled children. But what do we do if there isn't any hope?' Greta asks. 'Should we lie? Without action, hope sooner or later comes to an end – and then what'll we do? When the hope that everyone keeps talking about is no longer there? When another few years have passed and we still haven't started the gigantic readjustment that has to happen, and the hope we evidently can't manage without suddenly runs out? Should we just give up then? Lie down and die?'

Some cars drive past. An empty city bus rumbles on towards Bolinderplan and Kungsholmsgatan.

'And whose "hope" is it anyway?' she continues. 'What they call "hope" is as far from hopeful as it gets for me. Hope for me would be if the politicians called emergency meetings and there were big black headlines about the climate crisis all over the world.'

They walk their bikes down the steps to Kungsholms Strand and then ride home. Greta plops down on the couch with Moses and Roxy and watches animal videos on her phone.

Dogs are dancing on YouTube to a monotonous house beat. Greta laughs until she cries.

Scene 98.
Jay-Z

'Dancing is like breathing,' says Beata. And she dances all the time. Sometimes more than ten hours a day.

When she's not dancing, she's singing. Or acting. Her anger is gone. She finds purpose in all the artistic things she does, and that is enough to change everything around.

She has a strong will and enormous drive when she gets to do what she's good at.

In group-work at school, before the parliamentary election, they are shooting a film on a phone, and we get to see the clip of her rapping and talking right into the camera. It's a made-up commercial for a fictitious political party, but the result is astonishing. Her natural talent for acting is like a breath of fresh air.

We have never seen anything like it. Everything falls into place.

But overall, school is worse. There are new teachers every term. New substitutes every month. New classrooms.

And every week there is a new timetable to download on the City of Stockholm website. It is a procedure that for a computer-savvy person equipped with online ID takes five to ten minutes. For me it doesn't work at all.

Nor does it work for Beata.

It's as if the school deliberately creates an obstacle course to disadvantage anyone who likes routine.

Anyone who doesn't love constant change.

There are no limits to the number of impressions the children are supposed to experience every day.

Everything and everyone must perpetually be in constant motion.

Every week new things are planned, each and every one of which requires small outings around the city.

Field trips.

Visits.

Variation.

There is constant ongoing preparation for possible trips and exchanges, and – 'It's so amazing that the children get the opportunity to experience new places and maybe meet other children from faraway countries.'

As if those opportunities aren't available in any neighbouring European area.

But of course it must be the right sort of encounter. The right sort of place. The right sort of parents. And the right sort of children.

The worst thing isn't that this happens.

Nor is the worst thing that the schools know that many students are disadvantaged by the importance placed on the social competence that goes into creating the flexible, extroverted norm that typifies the idea of a successful student.

The worst thing is that many of the students know that this happens deliberately.

Especially those students who are hit the hardest.

They know.

They understand the betrayal.

They understand how their respective 'failures' are deliberately orchestrated to the advantage of the extroverted winners of the everyday.

*

Beata sits with Greta one day in front of Parliament.

But this is Greta's thing.

Not hers.

The sudden fuss over her big sister is not easy to handle.

Beata sees that Greta suddenly has 10,000 followers on Instagram, and we all think that's crazy.

But Beata handles it well.

Very well.

Even when her own feed is filled with comments about Greta, and can you tell her this and that. All everyone suddenly cares about is Greta, Greta and Greta.

'It's nuts,' Beata says one afternoon after school. 'It's exactly like Beyoncé and Jay-Z,' she states, with an acerbic emphasis.

'Greta is Beyoncé. And I'm Jay-Z.'

Scene 99. Crime against Humanity

Humanity is heading for a catastrophe. Every week new figures and reports appear, confirming that we are moving in the wrong direction.

At the highest possible speed.

And with each passing week the message from this research and science becomes clearer.

Everything becomes more starkly black and white.

We wonder who is most to blame.

Is it the oil and energy companies? The clothing companies, the fast-food chains? The forestry companies or industrial livestock producers? All of whom, within the limits of the law, do their utmost to sell as much as possible, so that they can maximize profits and returns for their shareholders.

Is it the politicians – who'll do anything to be re-elected?

Is it the newspapers – that must turn a profit in order to survive? Compelled to publish the sort of thing that everyone wants to read?

Is it us ordinary people? Who continue to consume with each passing day in order to get our increasingly unreasonable lives to function?

Is it me? Who had the opportunity to get involved in the situation but who chose to rely on politicians, businesses and the mass media?

Is it the scientists – who often lack the ability to communicate their findings? And whose knowledge of behavioural science has perhaps led them to talk of a crisis impending in twenty or thirty years – when it is suddenly happening here and now, sooner than anyone counted on? Much sooner than some of them anticipated it would?

Is it public service media – which is economically independent and supposed to keep an eye on every aspect of our society and its consequences for future generations, but whose employees are showered in hate by the ideological opponents of independent media? Which has been forced into the race for clicks and viewer numbers?

We are all to blame. But perhaps not equally.

Scene 100.
The Price of Being Heard is Hate

Even though Greta keeps repeating that the climate crisis can only be solved through democracy, she's constantly being accused of advocating a 'climate dictatorship'.

Even though she keeps repeating that there are no solutions within the prevailing political and economic systems, she is accused of not having any answers.

This is a deliberate strategy, of course.

Because it's not about listening and finding solutions. It has never been about that.

Because who wants a solution to a crisis which in their eyes doesn't exist? That cannot exist. Because if it did, wouldn't it mean that everything has to change?

Let's say the climate crisis is the existential crisis that the united science says it is. This would mean that the prevailing world order is responsible for a failure of cosmic proportions, since the threat is much greater than anything humankind has ever faced before.

No, that's inconceivable for anyone who does not want to see comprehensive change.

Better, then, to talk about law and order.

Or security.

Crime, refugees, jobs and money.

Always money.

Because what could possibly be going wrong when everything just keeps getting so much better, bigger, stronger, faster?

Nothing can be going seriously wrong, anyway.

Well, nothing except for the development of children.

Because, by the critics' logic, a fifteen-year-old can no longer think for herself, in spite of the fact that she is equipped with an unlimited data and is constantly connected to all the world's digital sources of knowledge.

You see, the children aren't following the general trajectory of a growth-oriented society. According to the critics of the school strike, in this case, the development is regressive.

Once fifteen-year-olds could be mothers, workers, soldiers, independent individuals, but nowadays they aren't good for anything. They can't even think a single thought for themselves.

And there are no exceptions – not as long as their thoughts aren't in line with those of certain adults, of course, and anyway: Shouldn't these youngsters be in school and learn some manners?

If they really want to save the world they should start by getting a proper diploma so that everything can follow its right and proper order. Then they can continue studying to be engineers and scientists so that in ten or fifteen years they can enter the working world and make a real difference.

The critics don't seem to want to absorb that it'll be too late by then.

Because for them this type of climate crisis – the type that requires action and change – does not exist. And herein, I guess, lies the ingenious nature of the school strike.

It is as simple and provocative as it needs to be.

The clock is ticking. Time is running away from us, and what can illustrate this better and more clearly than our own children's education?

What should they study to be?

And why?

The time we have left to act and to fundamentally change society is suddenly shorter than an average elementary and high-school education.

And when no sea-change is in the offing . . .

What should the children do then?

When they're being robbed of their most basic conditions for existence?

They can't vote, after all.

Much less can they influence industry, science, the mass media or political decisions.

Those who are most affected have no influence.

Our comfort is suddenly being set against their future.

All those things we *have to* get to do.

Our recreational interests against their survival.

Our growth at the expense of their world.

Our hobbies against their fundamental human rights.

It's tragic enough that for a long time we've been doing the exact same thing to people in poorer parts of the world.

But that argument doesn't sink its teeth in, apparently.

Because we obviously don't care.

Screw them.

On the other hand, the majority of us can't ignore our own children and grandchildren as easily.

The school strike seems to be working.

The tension between our excess and the inheritance we're going to leave to future generations creates exactly the friction and resistance needed to generate new debate and new emotions.

New angles.

To be sure, this is quite unintentional.

Because these things can't be planned.

They simply happen.

One attempt in a million.

Or perhaps a billion.

*

The striking children say that the solution to the crisis is to treat it as a crisis. Not exactly a new thought when it comes to the climate crisis – it's been out there for decades.

But it's not really about that. It has never really been about presenting new options or solutions.

It's about the vast majority of people wanting to carry on as usual.

Our human fear of change.

And the fact that that driving force happens to coincide with the preservation of the status quo, to the advantage of those who are the most privileged, is quite practical for those who happen to be part of that exclusive little group.

Their ability to engage so many angry, bitter, underpaid and exploited men to fight on their side is and will remain a fascinating phenomenon.

A kind of Catch-22 for humanity that perhaps isn't quite as mysterious as one would think.

Because if you're one of the winners in the prevailing world order, then naturally you'll go to great lengths to defend it. And what could be better than getting *the losers* in that prevailing world order to fight for the same cause?

'Losing' is, of course, relative and in this particular instance we are all more or less losers.

The recruitment base is practically unlimited and the secret is so simple. All you have to do is get as many people as possible to defend their little part of the universe.

Their job. Their home. Their holiday. Their car. Their money.

It's about scaring as many people as possible with the threat of change and decline. And doing it to such an extent that in principle they are prepared to do anything to stand up for their own microscopic part of this gigantic world.

To defend the prevailing food chains against anything and anyone that poses a threat to that stability.

Immigrants, refugees, liberals, socialists, feminists and activists.
The method is as simple as it is effective.
As brilliant as it is completely idiotic.

Greta provokes. In certain cases to such an extent that normally respectful people lose their composure. Not only does she say that everything has to change, she has autism too. And she has the gall to brag about it.
That's not how things are supposed to work.
The two things are incompatible – even if subconsciously – with the contempt for weakness of certain human ideologies and opinions.
They're incompatible with the competitive society's unwritten manifesto that says the strongest always win.
It is the strongest who will be heard.
It is the strongest who will set the agenda.
Such are the laws of the market, the jungle, the universe.
But on the cobblestones in front of the Swedish Parliament, suddenly other rules apply.
The invisible girl who never says anything is suddenly the one who is heard and seen the most. And apparently that is far too disturbing for everyone to let slide.
The hate grows stronger with each passing minute.
Stories, lies and personal attacks.
But the primary weapon of course is *the deliberate omission of facts*.

Greta's background and story are public knowledge on the internet and with a simple search you can read up on all the relevant, accepted facts. But what do those matter when the lie is much more entertaining? When the deliberate omission of facts gets more readers?

Scene 101.
The First Opening

The days go by, and suddenly Greta has been sitting there for two weeks.

Every morning she cycles off to Parliament and parks her bike by the iron railings in front of Rosenbad.

Every morning she meets the rest of us.

We who are busy with other things.

Sitting in our cars and listening to our radio programmes.

Standing with our mobile phones on the metro.

Sitting on buses and daydreaming.

Talking about the food we've eaten and the football we've watched.

Cleaning our houses and our apartments.

Washing our windows, arranging our pillows and sorting our bookshelves.

We who assume that everything is more or less as it should be.

The *Guardian* comes by and publishes the first major interview in the foreign media. Some Norwegian and Danish media have already written stories but this is next level.

Greta tells her story to anyone who asks. She answers all the questions and devotes what time is left to her books.

Everyone thinks that Greta's immense public journey started on the cobblestones outside the Swedish Parliament on 20 August 2018.

But that's not so.

It started much earlier.

I go back to an earlier post I made on Facebook. It has over 11,000 likes and Greta is being praised in hundreds of comments. She inspires hope and everyone seems to be taking in her words and thoughts. The post is nothing new. And it has nothing to do with her school strike.

It was written early on the morning of 9 November 2016 and it has never been edited.

That morning Stockholm was covered in over half a metre of fresh snow. A few hours earlier Svante had crawled off the couch to take cover on the floor because it felt as if 'an ice-cold wind passed through the apartment' when the election barometer suddenly turned from Hillary Clinton to Donald Trump.

That night, before the break of dawn, the USA had elected Donald Trump as their new president.

I wrote:

> Many people feel great fear this early morning. I am one of them. But we mustn't give in to the fear. We must stick together. Right and left. Across party divisions. We have to start a counter-movement here and now. We must organize ourselves against the darkness and the hate that has arisen in the ever-greater gaps that have emerged in the world. But we must never meet hatred, racism and bullying with the same hatred and bullying. We must never ever stoop to hate. We must start reducing the gaps instead. We must stand united for humanism and for the equal worth of all people. When they go low, we go high. Now is not the time to grieve or be afraid.
>
> Now is the time to organize.
>
> P.S. My elder daughter is passionate about the environment. She is much more well-read and knowledgeable than I am. This is what she's been saying:

'When the climate issue is as acute as it is now, probably the only salvation is that Donald Trump wins the election – because only then will people perhaps understand how bad it is. When a climate-change denier like Trump wins and becomes the world's most powerful man – then perhaps people will finally wake up and become sufficiently shaken to start the gigantic counter-movement that is needed for us to achieve a real change in time.'

Her words feel so incredibly hopeful and valuable today. I'm going to wake her up soon. With all the hope I have. It's time to start fighting. For her and for all our children.

Greta woke up with a smile that morning. She rubbed her eyes and looked up at the poster of the periodic table above her bed. But before she rattled through the elements, as she does right after she wakes up, she said, 'Obviously this is terrible. But it's the only way. With Clinton or Obama everything would have continued as before. Trump is the wake-up call.' It was a provocation from her side, of course. She knew very well that Trump's election was a disaster for everyday people. His rise to power exposed for many just how much things need to change.

I consider sharing the post again now during the school strike, but I stop myself.

Everything has its time, I think.

Let them hate their bloody hate now so everyone gets to see what kind of people they are.

Our family has been experiencing it for a long time.

We get death threats on social media, excrement through the letter box, and social services report that they have received a great number of complaints against us as Greta's parents. But at the same time they state in the letter that they 'do NOT intend to take any action'. We think of the capital letters as a little love note from an anonymous official at the Kungsholmen District Board. And it warms us.

But I can't ward off all the hate. Can't push it away from me. Because somehow I am starting to realize that they are going to take my child from me. She might not be able to keep on living here.

The price of being heard is hate.

The price of being seen is hate.

The price of everything is so terrifyingly much hate.

The hate knows no bounds.

And the haters are never going to stop hating.

Scene 102. Backward Steps

More and more people are keeping Greta company in front of Parliament. Children, adults, teachers, retirees.

The photographer Anders Hellberg is there every day. He photographs and posts pictures for all to use. He doesn't want a penny.

'Anyone who wants to can use the pictures. It's my way of trying to help out.'

One day an entire class of elementary-school pupils stops and wants to talk, and Greta has to walk away for a bit.

Feels mild panic.

She steps aside and starts crying.

She can't help it.

But after a while she calms herself down and goes back and greets the children.

Afterwards she explains that she has a hard time associating with children sometimes because she has had such bad experiences.

'I've never met a group of children that hasn't been mean to me. And wherever I've been I've been bullied because I'm different.'

It takes a toll to sit in front of Parliament seven hours a day for three weeks.

Lots of people want to come up and talk.

Usually these are pleasant people who want to be supportive and tell her that they have listened to what she is saying.

Several times a day people come up and say that they have stopped flying, parked the car or become vegans thanks to her.

To be able to influence so many people in such a short time is bewildering in a good way.

But of course you can't escape the critics.

Many want to have a discussion.

'What's the toughest one to have?' I ask.

It's Sunday, our day off, and we are sitting spread out on the living-room floor.

'There are a lot of different arguments,' Greta replies. 'Those who say that there are "too many people", for example. Partly because if some people say that there are too many of us, then I suppose they want us to get rid of some, because that's what the logic implies. And then you assume that it's either us children or people in developing countries who are the problem, because it's us children who are the last ones to arrive and spoil the party for all those who keep saying "there are too many people in India, Africa and China." But the fact is that the great majority of people on earth are not living above their means. It's those like us in Sweden who are. We're the ones who are living as if we had four globes, and at the same time think that there are "too many people". And if everyone were to live like we do, the two-degree target would have been lost long, long ago. Then there wouldn't be any future at all.'

Greta is sitting on the rug with Moses in front of her. He is sleeping, stretched out across the red patterned rug that we bought at an online auction almost ten years ago. No matter how much dirt and dog hair has collected on it, it always looks nice and clean.

'Then there are those who want to talk nuclear power,' she continues. 'They never talk about anything but nuclear power. It's as if there were no climate or ecology crisis at all. They just want to talk about nuclear power. They don't know any facts. They haven't even heard about the most basic things. They just say: "So what's your take on nuclear power?" And then they smile as if they have solved all the world's future problems on their own. But what's frightening is that politicians do the same thing. Because they know that nuclear power is no longer a solution. And yet they repeat the same thing.'

'What do the researchers say?' I ask.

'The IPCC says that nuclear power can be a small part of a big, holistic solution,' Svante answers. 'But they also say that the energy issue can be solved with renewable energy alone. It's not up to science to take a position, so the scientists are only talking about what is physically possible. Climate science doesn't neccessarily have to take politics and practical circumstances into account. The fact that in practice it takes ten to fifteen years to build a new nuclear reactor today – and that we would need thousands of completed plants tomorrow – is not something they need to consider, I guess.'

Roxy plops down on the rug alongside Greta and Moses. She licks her paws clean and stretches out, like a mirror image of Moses.

Within seconds, she's asleep.

'Okay, so we need a huge amount of new, fossil-free energy. And we need it now,' says Greta. 'So we must invest in the best, cheapest and fastest alternative. So why invest in something that takes over ten years to build when wind and solar can be ready within a few months? Why invest in something that is so expensive that no company wants to invest in it, when wind and solar are much cheaper, and falling in price by the minute? Why invest in a high-risk technology when you can invest in something risk-free? We haven't even solved the problem of storing existing nuclear waste. And if we were to replace all fossil energy with nuclear power we would need to basically complete a nuclear power plant a day, starting today. Just educating engineers to construct them would take decades. So nuclear power is pretty much an unfeasible alternative. Everyone knows that. So why do they keep talking about it?' she repeats. 'It really scares me. Because either the politicians are so stupid that they don't understand, or else they just want to waste time. And I don't know which is worse.'

'I think that the nuclear power question has enormous symbolism for a great many people,' Svante says. He has climbed up on the bar stool by the kitchen island that extends into the living room. 'If you don't want to talk about the climate, you can always talk about nuclear energy, because you know the conversation will get stuck there. Nuclear power is a climate delayers' best friend. I know, because I was like that myself. Only a few years ago, I thought it was a good solution to continue using nuclear power and there was something incredibly boring and backward-looking about all the environmentalists who wanted to shut it down. I think it has to do with optimism and the future. I wanted to believe that people could fix everything. That we had managed to find all the solutions. Because if we had succeeded, we wouldn't need to change. And so we wouldn't need to alter this prevailing world order that allowed someone like me to be able to travel basically anywhere I wanted to travel at almost any time. In that case I could buy that Range Rover I was secretly dreaming about. And I could eat whatever I wanted, because people had succeeded in taming nature and nothing needed to be changed. Except possibly having a bit more order.'

Svante scratches his head, stretches his back and spins halfway around on the stool before he continues.

'I think we should avoid talking about nuclear power. Because unless we're talking about holistic solutions it's uninteresting. Maybe it was different five or ten years ago. Then there was still a chance that a serious expansion of nuclear power could be part of the solution. But this is a different crisis now than only two or three years ago.'

'Why are certain politicians against wind and solar power?' I ask. 'Is it because it's cheap? Too simple? Because everyone can build their own energy system and countries and communities can become truly independent?'

Roxy wakes up. She gets up and noses around Greta and Moses before she lies down again. This time with her head against Moses' back leg.

We sit quietly a while. Under her chest you can see her little Labrador heart beating. Greta strokes her black fur and says, 'But the hardest of all are the people out to sell something. All the "Hi, I have a company and was wondering if you would consider collaborating with us." Or those who come up and want to invite me to various conferences, or who want to do a book, documentary or whatever. All the chancers. Those of us on school strike are saying that everyone should take a few steps *back* because it's the only way to save the climate, but then we're met by all those who want to move forward. Everyone looking for their opportunity, wanting to invest in themselves and become someone they aren't already.'

Seven billion people, all of whom want to realize themselves, I think. But in fact it's not true.

Only a small minority lives outside the planetary limits of what is sustainable.

The problem is that we belong to that minority.

The problem is that we who already have enough in every way are encouraged to take more.

Buy more.

Drive more.

Eat more.

Do more.

★

Sometimes we think about how it was before all this.

How was it possible that we didn't see what we so clearly see today?

And what would we have looked like if it hadn't been for our daughters?

Would we have continued as usual the past three or four years had it not been for them?

What would our everyday existence have been like if we hadn't acknowledged our own failures the moment our arguments ran out?

I want to believe we would have acted anyway. That we would have changed our lives.

But I doubt it.

Sometimes we think about how we would have reacted if suddenly we saw a little fifteen-year-old girl sitting outside Parliament 'on school strike for the climate'.

Would we have chosen not to listen to her?

Would we have closed our eyes?

Would we perhaps have chosen to accept one of the conspiracy theories because *it must be something shady*?

Would we have blamed everything on China?

Would we have been upset by the striking girl?

Even hated her?

Would we have chosen to look the other way so that we could go on like before?

Would we – honestly – have chosen to take a few steps back of our own free will?

Scene 103. Dress Rehearsal

The phenomenon keeps growing. Faster and faster by the hour. In the run-up to the end of the strike, Greta is being followed by TV crews from the BBC, German ARD and Danish TV2.

I have dress rehearsals in the evenings. It will soon be time for the premiere of the musical *As It is in Heaven*, and workdays in the theatre are long. Greta is asleep when I come home and I'm asleep in the morning when she leaves. I don't hear the TV crews sneaking around in the apartment, filming Greta's morning routine.

When the last Friday arrives there are strikes in over a hundred places all around Sweden. In Germany, Finland and Great Britain some scattered individuals have also joined in. In the Netherlands a hundred children are striking outside Parliament in The Hague. And in Norway there are several thousand.

It feels dizzyingly big.

Janine O'Keefe is one of the activists who has joined the strike and she is trying to organize everything. She is from Australia and has a small network of other activists she has known for a long time. Fältbiologerna and Greenpeace also help out, as do Klimatsverige, Naturskyddsföreningen, We Don't Have Time, Stormvarning, Föräldravrålet and Artister För Miljön.

Anyone who in some way is fighting for the environment and the climate is helping out in their own way.

As much as they can.

Altogether a thousand children and adults sit with Greta on

the last day of the school strike. And media from several different countries report live from Mynttorget Square.

She has succeeded.

Greta has carried out her plan.

She has spent three weeks striking outside the Swedish Parliament.

She has seen to it that the climate issue is a little more in focus.

Or quite a bit more, really.

Some say that she alone has done more for the climate than politicians and the mass media have in years.

But Greta doesn't agree.

'Nothing has changed,' she says. 'The emissions continue to increase and there is no change in sight.'

At three o'clock Svante comes and picks her up and they walk together through the arch on Riksgatan over to the bicycles outside Rosenbad.

'Are you satisfied?' Svante asks.

Greta is silent.

He repeats the question, but Greta does not reply.

They unlock the bikes and get ready to cycle home.

'No,' she says, with her gaze fixed on the bridge back towards the Old Town. 'I'm going to continue.'

Scene 104.
Fridays for Future

The next day is Saturday 8 September. It's the day before the Swedish parliamentary elections and Greta is going to speak at the People's Climate March in Stockholm. Around the world, tens of thousands of people are going to march for the climate. Many hoped for a huge global demonstration, but it is doubtful there is enough interest.

Many are still hoping, but despite the summer's fires and the increasingly extreme weather all around the world, things are moving sluggishly for the international climate and environmental movement.

Greta will speak at the end of the march, up by the palace. It was booked long ago. She intends to read a piece she wrote for the newspaper *ETC*.

But now she wants to give an additional speech.

At the beginning.

Before they set off.

Svante questions if it's a good idea.

She has only given one speech before. It was on Nytorget, outside a restaurant where some performers, several of our friends, were encouraged to 'back Greta' at a support concert.

Until then she'd never spoken in front of more people than fit in a classroom, and on those few occasions she had not exactly seemed at ease.

On the contrary.

But she is stubborn, and Svante calls Ivan at Greenpeace, and Ivan says that it's complicated, there being so many competing interests ahead of the demonstration and all, but he'll arrange it anyway.

'Somehow.'

There are a lot of people in Rålambshovsparken. Almost 2,000 have crowded together at the Park Theatre's stage behind the green hills up towards Västerbron. There are already twice as many as usually come to climate demonstrations. And more are on their way.

The air is mild.

The trees, streamers and banners move with the wind, and although everyone knows that this is nowhere near what is required to put the climate issue front and centre, there's a different feeling about this protest.

It doesn't feel the same as usual.

It feels as if something might happen.

Soon.

Perhaps it's the mixture of people.

It's no longer just the familiar faces. The regulars. The activists. The Greenpeace volunteers in polar-bear suits.

Here, suddenly, are all conceivable kinds of people and characters.

People who might have all sorts of jobs. Whose votes might go any which way.

'This is my first demonstration,' states a well-dressed man in his forties.

'Mine too,' a woman next to him says, with a laugh.

The host introduces Greta and she walks slowly but steadily into the middle of the amphitheatre's gravel stage. She is accompanied by three of the girls who have gone on strike with her for the past two weeks: Edit, Mina and Morrigan.

The audience cheers.

Svante, on the other hand, is scared out of his wits. What will happen now?

Is she going to speak? Will she start crying? Is she going to run away?

He feels like an awful parent for not putting his foot down and saying 'No' from the start. All this is starting to feel too big and unreal.

But Greta is as calm as can be.

She takes the speech out of her pocket and looks out over the fan-shaped gallery. She lets her gaze wander over the sea of people.

Then she grasps the microphone and starts speaking.

'Hi, my name is Greta,' she says in Swedish. 'I am going to speak in English now. And I want you to take out your phones and film what I'm saying. Then you can post it on your social media.'

The audience gives a surprised chuckle, and people take out their phones and get ready to film. For a few seconds almost everyone has their phones aimed at the four teenagers on the stage.

'My name is Greta Thunberg and I am fifteen years old. And this is Mina, Morrigan and Edit, and we have school-striked for the climate for the last three weeks. Yesterday was the last day. But . . .'

She pauses.

'We will go on with the school strike. Every Friday, as from now, we will sit outside the Swedish Parliament until Sweden is in line with the Paris Agreement.'

The crowd cheers.

Many people have told her that the strike must have a list of demands to submit to the politicians. A manifesto or something.

But Greta refuses to make any specific demands.

As she has explained again and again, 'If we propose a lot of specific solutions, everyone is going to think that is enough. But it won't be. We need system changes and a new way of thinking. What has to happen – what is written in between the lines in all the agreements and reports – is so much more radical than any manifesto could ever include. Our only chance is to turn over all that must be done to the scientists. We are children. We can only refer to what the scientists say.'

The gentle late-summer breeze plays in the treetops, high above Rålambshovsparken. The cheering dies down and Greta continues.

'We urge all of you to do the same. Sit outside your parliament or local government, wherever you are, until your country is on a safe pathway to a below-two-degree warming target. Time is much shorter than we think. Failure means disaster.'

Greta has the microphone in her right hand and in her left the folded-up paper from which she is reading. Her voice is steady and there are no signs of nervousness. She appears to be at ease up there. She even smiles sometimes, and in the stands Svante has already calmed down.

'The changes required are enormous and we must all contribute in every part of our everyday life. Especially us in the rich countries, where no nation is doing nearly enough. The grown-ups have failed us and since most of them, including the press and the politicians, keep ignoring the situation we must take action into our own hands. Starting today.

'Everyone is welcome. Everyone is needed. Please join in. Thank you.'

The audience stands up. Shouting, applauding.

'You must be very proud,' says the woman next to Svante. She recognizes that he's Greta's dad.

'Proud?' Svante repeats. Loudly, so he can be heard above the cheers. 'No, I'm not proud. I'm just so endlessly happy because I can see that she's feeling good.'

The ovation doesn't stop. Greta leans over to Edit and whispers. They nod at each other.

And Greta is smiling the most beautiful smile I have ever seen her smile.

I'm watching everything from a live stream on my phone in the hallway outside the dressing rooms at Oscarsteatern.

The tears keep coming.

Scene 105. Hope

The question is how we want to be remembered.

Those of us who lived during the time of the great fires.

What are we going to leave behind us?

From a sustainability perspective, we have, so far, failed on every point.

But.

We can change all that.

And we can do it very fast.

We still have a chance to put everything right, and there is nothing we as people can't achieve, if only we want to.

Hope is literally everywhere, but that hopefulness makes demands.

Because without demands hope is hollow. Without demands hope is simply standing in the way of the major change that is required.

My hope affirms our good intentions and our short-comings.

The way forward is not via prodding or witch hunts; it isn't about pitting individual actions against each other.

My hopefulness demands radical action.

My hopefulness is not talking about what others should do or about what we can manage in ten years, because in ten years it may already be too late.

My hope is always right here and now, and I am convinced that the politician who chooses to advocate radical change is in for a positive surprise. If he or she is prepared to try to live as he or she advocates.

Humanity's greatest leaders, the ones who will be remembered, have all had one thing in common – at the right moment they have chosen to put our future ahead of the present.

And if our fate now rests in the media's hands, it could not have found itself in a better place.

The media know what responsibility rests on their shoulders. They know what editorial choices have been made and how they can be corrected. They know their reputations are at stake.

Each individual action is part of a common movement that is growing stronger and bigger every day. While waiting for the role models, the news desks and politicians, we have to do all the things we can do.

And all the things we can't do.

We have to abandon the maps and make our way into the unknown.

We have to start listening to all that we've stopped listening to.

We have to take the lead while keeping a welcoming door open behind us.

Everyone is welcome.

Everything and everyone is needed.

Scene 106.
Back to Square One

Late one evening when the apartment is in darkness, the phone chimes. Greta and Svante and the dogs went to sleep long ago. It's Beata, texting from her bed on the top floor.

'This is so me,' she writes.

Beata has sent a YouTube link and a screen dump from a website. 'Misophonia', it says.

'I was searching for diagnoses,' she says. 'This describes exactly how it feels.'

I read. I scroll down.

Read a little more. What is this? Yet another dead end? Another side track created by querulants hoping to make a career in having as many diseases and strange conditions as possible?

But no.

Apparently misophonia really exists. Articles have been published in TIME magazine, the *New York Times* and lots of other newspapers, and it all fits with Beata.

Literally everything.

Misophonia is a neurological syndrome, a disorder, that involves intense discomfort with certain, specific sounds. Everyday sounds. For example, breathing in and out, lip-smacking, whispering. Or tableware against porcelain.

Of course everyone is bothered by certain specific sounds. But for a person with misophonia these so-called trigger sounds are often so disturbing that they simply can't function in various contexts. The most common reaction is anger and rage.

*

For example, Beata has told us again and again how she can't concentrate on anything if she hears someone whispering.

'It can't be controlled. When you're sitting next to someone who snuffles through their nose you can't do anything about the fact that you get angry. It's impossible to focus on anything else.'

Misophonia is a brand new concept but it exists, with the relevant supporting research and everything.

A study from Amsterdam University recommends that misophonia immediately become a recognized diagnosis, because those who are affected have a clear and obvious handicap that they can't control.

A Newcastle University study published in 2017 says that 'Misophonia [is] a devastating disorder for sufferers and their families, and yet nothing is known about the underlying mechanism.'

There are connections to ADHD and autism spectrum disorder. There are connections to stress. Sometimes the symptoms go away as you grow older. It's all very real.

And yet I've never heard the word mentioned. Despite all the thousands of pages I've read. Despite all the meetings and conversations that I've attended.

'The awareness of misophonia is reminiscent of how ADHD was viewed just a couple of decades ago,' an American psychologist and author writes.

There are resources. Adaptations can be made.

But here there are no maps yet. Everything is uncharted territory.

And so we're back to square one. Again.

Scene 107. Safety Valves

There is so much we don't know. Many people say that we humans have already understood the significance of the climate and sustainability crisis. And we're repressing it.

But that's incorrect.

Our ignorance is so much greater than we think.

We don't know.

We don't understand.

Decades of important discoveries have not got through. The information has almost been for nothing.

A crushing majority of the world's population has no idea about the true meaning of the climate crisis.

And right there, in that knowledge, is all the hope we need.

Because what if we'd known?

What if we'd done all that we've done maliciously? Completely deliberately?

What if today we were to keep on doing what we do, despite our understanding of the ecological catastrophe we are leaving behind us?

As if humans were deliberately evil?

That is inconceivable.

And what if humanity's pain threshold had been a little higher?

What if we could have continued to live like we do without so many of us starting to fall apart and break down in the frenzy of society today?

Then it all would have been too late.

Then all social injustice, all oppression, all mental illness and all burn-out would have been in vain.

But it's not like that.

It turns out there are a few safety valves, and this is one of them.

It says that there is still time.

It says that there is a political system that gives us the chance to repair what has been broken and create something just, new and better. It says that there are tools and those tools are called public awareness and education.

The climate crisis is one symptom among many of an unsustainable world.

An acute symptom.

At the same time the sustainability crisis is a choice.

A chance to put everything right.

And in that lies our hope.

Scene 108. Places, Everyone

I believe that life is not a dress rehearsal. And I believe that we have to rethink our approach.

Of all the people who have ever populated this earth, 7 per cent are alive today.

That's us.

We belong together. We are a part of a whole that extends backwards and forwards in time and it's up to us, this 7 per cent, to secure everyone's future.

That is our historic task and we need each other.

More than ever.

We need technology. We need sustainable forestry and agriculture. We need the companies, the economists, the politicians, the journalists and the scientists, and we need our amazing capacity for adaptation and change.

But most of all we need to affirm each other's goodwill.

We have already solved the climate crisis. What needs to be done is crystal clear.

All that's left to do is make a choice.

Economy or ecology?

We have to choose.

At least until we have made our way to safe ground.

The fact that our purely existential challenges are still being reduced to party politics is absurd. To secure the limited resources that enable future life must be an absolute given. Just like the insight that the way forward can sometimes require taking a few steps back.

Just like equality and the fact that every person's equal worth ought to be as self-evident as the political parties maintain it is.

But it isn't.

On the contrary.

And for that very reason there are no questions that are more political.

They belong together.

They are one and the same.

Because when our limitless, hyper-competitive society's CO_2 levels reach our outermost atmosphere and literally hits the ceiling, when the law that everything must get bigger, faster, greater is set against our common survival, a new world stands at the door, and that world has never been as close as it is now.

Or as far away.

A moderate world.

A world where a little girl equipped with an Instagram account and a picture of a polar bear can be an equally effective defender of our common security as all the world's armies combined.

Our limitations are slowly becoming apparent. The infinite is regaining its contours. Not everything is possible, and that's good. Because in moderation there is a different, much greater freedom.

Is the struggle for the environment the world's greatest feminist movement? Not because it in any way excludes men, but because it challenges the structures and values that have created the crisis we now find ourselves in.

Mother Earth is at the ready backstage.

At any moment the curtain will rise.

We have to start talking about how we're doing.

Because it's up to us now.

It's us against the darkness.

From mouth to mouth, from city to city, from nation to nation.

Organize.

Act.

Make ripples on the water.

It's places, everyone.

It's time to take your place on the stage.

Thanks for Your Help, Energy and Inspiration

Anita and Janne von Berens, Anna Melin, Camilla Berntsdotter Lindblom, Hanna Askered and Lära Med Djur, Björn Meder, Jiang Millington and Barn i Behov, Pär Holmgren.

Nils Erik Svedlund and the Stockholm Centre for Eating Disorders, Svenny Kopp, Kevin Anderson, Isak Stoddard, the staff at BUP Kungsholmen, Magdalena Mattson, Kerstin Avemo, Fredrik Kempe, Lina Martinsson, Helen Sjöholm and David Granditsky, Jenny Stiernstedt, Hundar Utan Hem, Leif Blixten Henriksson, Pernilla Thagaard, Stefan Sundström, Mårten Aglander, Jonas Gardell and Mark Levengood, Mina Dennert, Mats Bergström, Janne Bengtsson, Petronella Nettermalm, Sten Collander, Ola Ilstedt, Stina Wollter, Anders Wijkman, Özz Nûjen, Fredrik Marcus, Karin af Klintberg, Johan Ehrenberg, Alexandra Pascalidou, Staffan Lindberg, Björn Ferry, Heidi Andersson, Maja Hellsing, Jeanette Andersson, Mattias Goldman, Helle Klein, Nisse Landgren, Vicky von der Lancken, Kent Wisti, Anna Takanen, Cecilia Ekebjär, Rosanna Endre and Greenpeace Sverige, the staff at Oatly Sverige, Martin Hedberg, Malin Tärnström, Hanna Friman, Christoffer Hörnell, Susanna Jankovic, Tomas Törnkvist, Frida Boisen, Carl Schlyter, Rebecka Le Moine, Svenska Stråkensemblen, Oskar Johansson, Anders Amrén, Peter Edding, Helena Lex Norling, Djurens Rätt, Vi Står Inte Ut, WWF Sverige, Naturskyddsföreningen and our families.

An extra big thank you to Jonas Axelsson, Annie Murphy and everyone at Bokförlaget Polaris. And of course thank you to Elias Våhlund and Tom Goren, and to Sirkka Persson and the Staff at Kringlaskolan in Södertälje.